AF347530

The 2nd International Electronic Conference on Biosensors

The 2nd International Electronic Conference on Biosensors

Editors

Giovanna Marrazza
Sara Tombelli

MDPI • Basel • Beijing • Wuhan • Barcelona • Belgrade • Manchester • Tokyo • Cluj • Tianjin

Editors
Giovanna Marrazza
University of Florence
Italy

Sara Tombelli
Istituto di Fisica Applicata
Nello Carrara (IFAC)
Italy

Editorial Office
MDPI
St. Alban-Anlage 66
4052 Basel, Switzerland

This is a reprint of proceeding papers published online in the open access journal *Engineering Proceedings* (ISSN 2673-4591) (available at: https://www.mdpi.com/2673-4591/16/1).

For citation purposes, cite each article independently as indicated on the article page online and as indicated below:

LastName, A.A.; LastName, B.B.; LastName, C.C. Article Title. *Journal Name* **Year**, *Volume Number*, Page Range.

ISBN 978-3-0365-4603-2 (Hbk)
ISBN 978-3-0365-4604-9 (PDF)

© 2022 by the authors. Articles in this book are Open Access and distributed under the Creative Commons Attribution (CC BY) license, which allows users to download, copy and build upon published articles, as long as the author and publisher are properly credited, which ensures maximum dissemination and a wider impact of our publications.
The book as a whole is distributed by MDPI under the terms and conditions of the Creative Commons license CC BY-NC-ND.

Contents

About the Editors . . . . **vii**

Preface to "The 2nd International Electronic Conference on Biosensors" . . . . **ix**

Zheyuan Chen, Ting-Yen Chi and Jun Kameoka
Cancer Biomarker Methylmalonic Acid Detection by Molecularly Imprinted Polyaniline
Paper Sensor
Reprinted from: *Eng. Proc.* **2022**, *16*, 1, doi:10.3390/IECB2022-12250 . . . . **1**

Rassen Boukraa, Giorgio Mattana, Nicolas Battaglini and Benoit Piro
A Complementary Reduced Graphene Oxide-Based Inverter for Ion Sensing
Reprinted from: *Eng. Proc.* **2022**, *16*, 2, doi:10.3390/IECB2022-12272 . . . . **7**

Nataliya Stasyuk, Olha Demkiv, Galina Gayda, Roman Serkiz, Andriy Zakalskiy,
Oksana Zakalska, Halyna Klepach, Galeb Al-Maali, Nina Bisko and Mykhailo Gonchar
Highly Sensitive Amperometric Biosensors Based on Oxidases and CuCe Nanoparticles
Coupled with Porous Gold
Reprinted from: *Eng. Proc.* **2022**, *16*, 3, doi:10.3390/IECB2022-12251 . . . . **11**

Sofija S. Bekić, Ivana Kuzminac, Srđan Bjedov, Jovana Ajduković, Marina Savić,
Edward Petri and Anđelka Ćelić
Use of Fluorescent Yeast-Based Biosensors for Evaluation of the Binding Affinities of New
Steroid Hormone and Bile Acid Derivatives for Select Steroid Receptors
Reprinted from: *Eng. Proc.* **2022**, *16*, 4, doi:10.3390/IECB2022-12282 . . . . **19**

Michailia Angelopoulou, Panagiota Petrou, Konstantinos Misiakos,
Ioannis Raptis and Sotirios Kakabakos
Simultaneous Detection of *Salmonella typhimurium* and *Escherichia coli* O157:H7 in Drinking
Water with Mach–Zehnder Interferometers Monolithically Integrated on Silicon Chips
Reprinted from: *Eng. Proc.* **2022**, *16*, 5, doi:10.3390/IECB2022-12269 . . . . **21**

Ivan Piovarci, Judit Süle, Michailia Angelopoulou, Panagiota Petrou, Leda Bousiakou,
Sotirios Elias Kakabakos and Tibor Hianik
Application of Optical and Acoustic Methods for the Detection of Bacterial Pathogens Using
DNA Aptamers as Receptors
Reprinted from: *Eng. Proc.* **2022**, *16*, 6, doi:10.3390/IECB2022-12268 . . . . **29**

Sarra Takita, Alexei Nabok, Anna Lishchuk, Magdi H. Mussa and David Smith
Detection of Prostate Cancer Biomarker PCA3 with Electrochemical Apta-Sensor
Reprinted from: *Eng. Proc.* **2022**, *16*, 8, doi:10.3390/IECB2022-12257 . . . . **37**

Antonios Georgas, Konstantinos Agiannis, Vasiliki Papakosta, Spyridon Angelopoulos,
Evangelos Hristoforou and Angelo Ferraro
A Portable Screening Device for SARS-CoV-2 with Smartphone Readout
Reprinted from: *Eng. Proc.* **2022**, *16*, 7, doi:10.3390/IECB2022-12274 . . . . **45**

Uriel Abe Contardi, Mateus Morikawa, Bruno Brunelli and Douglas Vieira Thomaz
MAX30102 Photometric Biosensor Coupled to ESP32-Webserver Capabilities for Continuous
Point of Care Oxygen Saturation and Heartrate Monitoring
Reprinted from: *Eng. Proc.* **2022**, *16*, 9, doi:10.3390/IECB2022-11114 . . . . **53**

Nunzio Cennamo, Maria Pesavento, Daniele Merli, Antonella Profumo, Luigi Zeni and
Giancarla Alberti
An Optical Fiber Sensor System for Uranium Detection in Water
Reprinted from: *Eng. Proc.* **2022**, *16*, 10, doi:10.3390/IECB2022-12296 . . . . **59**

Ahmed Fatimi
Trends and Recent Patents on Cellulose-Based Biosensors
Reprinted from: *Eng. Proc.* **2022**, *16*, 12, doi:10.3390/IECB2022-12253 . . . . **65**

Panagiota Petrou, Eleni Makarona, Ioannis Raptis, Sotirios Kakabakos and
Konstantinos Misiakos
Monolithically Integrated Label-Free Optical Immunosensors
Reprinted from: *Eng. Proc.* **2022**, *16*, 11, doi:10.3390/IECB2022-12283 . . . . **75**

Florine Joosten, Marc Parrilla and Karolien De Wael
Electrochemical Detection of Cocaine in Authentic Oral Fluid
Reprinted from: *Eng. Proc.* **2022**, *16*, 13, doi:10.3390/IECB2022-12284 . . . . **87**

Nelson Arturo Manrique-Rodriguez, Sabrina Di Masi and Cosimino Malitesta
Development of Electrochemical Sensors Based on Electrosynthesized Imprinted Polymers for
Cobalt (Co^{2+}) Ion Determination in Water
Reprinted from: *Eng. Proc.* **2022**, *16*, 15, doi:10.3390/IECB2022-12281 . . . . **95**

Chrysoula-Evangelia Karachaliou, Georgios Koukouvinos, Katerina Pissaridi,
Dimitris Ladikos, Dimitris Goustouridis, Ioannis Raptis, Evangelia Livaniou,
Sotirios Kakabakos and Panagiota Petrou
Fast and Accurate Determination of Minute Ochratoxin A Levels in Cereal Flours: Towards
Application at the Field
Reprinted from: *Eng. Proc.* **2022**, *16*, 14, doi:10.3390/IECB2022-12270 . . . . **97**

Sotiria D. Psoma, Ihor Sobianin and Antonios Tourlidakis
Computational and Experimental Investigation of Microfluidic Chamber Designs for DNA
Biosensors
Reprinted from: *Eng. Proc.* **2022**, *16*, 16, doi:10.3390/IECB2022-12252 . . . . **103**

Nikitas Melios, Vasiliki Tsouti, Stavros Chatzandroulis and George Tsekenis
Development of an All-Carbon Electrochemical Biosensor on a Flexible Substrate for the
Sensitive Detection of Glucose
Reprinted from: *Eng. Proc.* **2022**, *16*, 17, doi:10.3390/IECB2022-12273 . . . . **117**

About the Editors

Giovanna Marrazza

Prof. Dr. Giovanna Marrazza is a full professor in Analytical Chemistry at the Department of Chemistry "Ugo Schiff" of the University of Florence, Italy. She was President of the Course of Study in Chemistry Degree, at this University from 2017 to 2021. She has been a distinguished Visiting Professor of the Faculty of Pharmacy "Iuliu Haţieganu" University of Medicine & Pharmacy, Cluj-Napoca (Romania) since 2017. She is a member of the steering board of Italian Sensors and Microsystems Group (AISEM). She is a member of the Italian Sensor Group, Italian Chemistry Society (SCI). Her research is focused on new biosensing principles containing nanomaterials and modified interfaces with nucleic acids, enzymes, antibodies, bacteria, and molecular imprinted polymers. She is an expert in the design of procedures that are suitable for biosensor devices, such as microflow systems, thick-film technology, and nano-dispensing technologies. She has contributed more than 139 papers to leading international journals and book chapters (H-index=42, Scopus). She is a unit leader in national and international projects; moreover, she has participated in several national and international projects.

Sara Tombelli

Dr Sara Tombelli is a senior researcher at the Institute of Applied Physics (CNR) in Florence, Italy. Her research activity is focused on analytical chemistry, biosensor development and surface modifications with bio-molecules, such as antibodies, enzymes, aptamers, and nucleic acid probes. She has experience in intracellular nanosensors and nanoparticle manipulation, as well as in the design and application of optical nanoprobes such as molecular beacons and oligonucleotidic optical switches. She has contributed more than 130 publications on the above topics to international refereed journals, books, and conference proceedings (H-index=38, Scopus). She is a member of the Editorial Board of Sensors; she has been a Special Issue Editor for several MDPI journals and for the Elsevier journal Sensors & and Actuators Reports. She has been part of international commissions as an international expert for several international PhD theses (Spain, Switzerland, Italy). She has participated in several European Projects and other national and international projects; she has been a unit leader in national projects and was responsible for several measurement campaigns in the frame of European projects.

Preface to "The 2nd International Electronic Conference on Biosensors"

This book contains 17 papers presented at the 2nd International Electronic Conference on Biosensors. The conference was held from the 14th to 18th of February 2022. The Conference offered a unique opportunity to virtually meet with international experts in the field of biosensors. This book of proceedings offers a widespread overview of the research status and development of biosensors in the world.

The advancements in the biosensors research is highly interdisciplinary, producing innovative results in different areas such as material science, chemistry, medicine, biology, microelectronics.

These were days of intense activity and involvement, which saw fruitful discussion and an interesting exchange of ideas with a distinctly professional value.

We would like to thank all those who contributed to the success of this conference.

Giovanna Marrazza and Sara Tombelli
Editors

Proceeding Paper

Cancer Biomarker Methylmalonic Acid Detection by Molecularly Imprinted Polyaniline Paper Sensor [†]

Zheyuan Chen [1], Ting-Yen Chi [2] and Jun Kameoka [1,2,*]

[1] Department of Electrical and Computer Engineering, Texas A & M University, College Station, TX 77843, USA; zychen@tamu.edu
[2] Department of Materials Science and Engineering, Texas A & M University, College Station, TX 77843, USA; kevin0149@tamu.edu
[*] Correspondence: kameoka@tamu.edu
[†] Presented at the 2nd International Electronic Conference on Biosensors, 14–18 February 2022; Available online: https://sciforum.net/event/IECB2022.

Abstract: Methylmalonic acid (MMA) plays a vital role in metabolism and energy production. It has been studied and reported as a sensitive early indicator for mild or serious Vitamin B12 deficiency. The normal range in health people is from 0.00 to 0.40 μM. Thus, most of MMA detection research was focused on Vitamin B12 deficiency with a small detection range. Recently, MMA has been reported to promote tumor progression due to age-induced accumulation. It was found that MMA concentration can reach as high as 80 μM in elderly people. MMA can be of great value as a promising biomarker for cancer diagnostics, as well as a therapeutic target for cancer treatment. Clinical determination of MMA concentration is by the method of gas chromatography mass spectroscopy (GCMS) or liquid chromatography mass spectroscopy (LCMS). However, these methods require extensive sample pre-treatment and large sample volume. They are also expensive and time-consuming. Hence, we proposed an attractive and effective strategy to detect MMA with a broad linear range by a low-cost molecularly imprinted polyaniline paper sensor. The polyaniline paper strip was fabricated by a one-step solution process using MMA as the template by molecular imprinting technology. The concentration of MMA was determined by the resistance change of the paper sensor. A calibration curve as a function of MMA concentration in aqueous solution was acquired with a correlation coefficient of 0.962. We demonstrated detection of the added MMA in plasma with a wide concentration range of 0 to 100 μM with a limit of detection (LoD) of 0.197 μM. This low-cost disposable paper sensor shows great potential in point-of-care MMA detection for cancer prognostics and diagnostics, especially in underserved communities.

Keywords: methylmalonic acid detection; cancer biomarker; molecular imprinting; polyaniline; paper sensor

Citation: Chen, Z.; Chi, T.-Y.; Kameoka, J. Cancer Biomarker Methylmalonic Acid Detection by Molecularly Imprinted Polyaniline Paper Sensor. *Eng. Proc.* **2022**, *16*, 1. https://doi.org/10.3390/IECB2022-12250

Academic Editors: Giovanna Marrazza and Sara Tombelli

Published: 14 February 2022

Publisher's Note: MDPI stays neutral with regard to jurisdictional claims in published maps and institutional affiliations.

Copyright: © 2022 by the authors. Licensee MDPI, Basel, Switzerland. This article is an open access article distributed under the terms and conditions of the Creative Commons Attribution (CC BY) license (https://creativecommons.org/licenses/by/4.0/).

1. Introduction

Methylmalonic acid (MMA) plays a vital role in metabolism and energy production. It has been studied and reported that the concentration of MMA increases in blood and urine if adequate amount of vitamin B12 is not available in the body [1,2]. Therefore, MMA is viewed as a specific functional metabolic marker, which is a sensitive early indicator for mild or serious vitamin B12 deficiency. The normal range in health people is from 0.00 to 0.40 μM [3]. Hence, most research in the field of MMA detection were focused on Vitamin B12 deficiency with a small dynamic range. Recently, MMA has been reported to promote tumor progression due to age-induced accumulation [4]. It was found that MMA concentration can reach as high as 80 μM in elderly people. MMA can be of great value as a promising biomarker for cancer diagnostics, as well as a therapeutic target for cancer treatment. Thus, it is quite urgent to investigate detection of MMA with a wide detection range, especially at high concentration conditions.

Many methods have been developed to detect MMA. Clinical determination of MMA concentration is by the method of gas chromatography mass spectroscopy (GCMS) or liquid chromatography mass spectroscopy (LCMS) [5,6]. However, these methods require extensive sample pre-treatment and large sample volume. They are also expensive and time-consuming. Electrochemical sensing has been reported as one of the reliable methods for determination of biomolecules [7,8]. However, this approach is limited by several issues. One of the major issues during sensing the biomolecules is that they tend to undergo oxidation/reduction at the closed potentials of the conventional electrodes. Furthermore, the electrodes have a fouling effect which negatively affects the selectivity and reproducibility. These kinds of shortcomings can be improved by modifying the electrodes that play a vital role in the effective detection of biomolecules [9].

Molecular imprinting (MIP) is a promising process that works by the co-polymerization of functional monomers and cross-linking agents in the presence of a template molecule or its derivative [10,11]. The MIP approach has shown great advantages, including simple preparation, potential reusability, and long-term stability [12]. MIP-based methods have been demonstrated as promising approaches to various chemical/bio sensing applications [13–17]. PdAu-polypyrrole tailored carbon fiber paper electrode has been reported to detect MMA, but it is limited by a narrow dynamic range, only from 4.01 pM to 52.5 nM [18]. A molecularly imprinted polymer modified with graphene oxide and gold nanoparticles was also reported to detect MMA with a small linear range (<4 µM) [19]. However, these sensors have narrow detection ranges. Furthermore, they used gold-modified electrodes which will drastically increase the cost for each sensor.

Hence, we proposed an attractive and effective strategy to detect MMA with a broad linear range by a low-cost MIP polyaniline (PANI) paper sensor. The concentration of MMA was determined by the resistance change of the paper sensor. A calibration curve as a function of MMA concentration in aqueous solution was acquired with a correlation coefficient of 0.962. We also demonstrated detection of the added MMA in plasma with a wide concentration range of 0 to 100 µM. This low-cost disposable paper sensor shows great potential in point-of-care MMA detection for cancer prognostics and diagnostics.

2. Methods

2.1. Materials

MMA powder (99%) was purchased from Sigma-Aldrich (St. Louis, MO, USA). Aniline (ACS reagent) and ammonium persulfate (APS, ACS reagent) were also purchased from Sigma-Aldrich (St. Louis, MO, USA). Hydrochloric acid (HCl, 36–38% w/v) and acetic acid (ACS reagent) were purchased from Macron (Center Valley, PA, USA). Polyester paper substrates were obtained from Xerox (Norwalk, CT, USA). Silver conductive ink was purchased from Creative Materials (Ayer, MA, USA). Human plasma derived from whole blood donations was purchased from BioChemed Services (Winchester, VA, USA).

2.2. Synthesis of MMA Imprinted PANI on Paper Substrates

The synthesis process using MMA as the template was modified based on the fabrication process of molecularly imprinted polyaniline (MIP-PANI) paper sensor reported previously [20,21]. As control, the non-molecularly imprinted (NIP) PANI paper strips were synthesized following the same protocol without the addition of MMA template into the monomer solution.

2.3. Fabrication of PANI Paper Sensor

The as-prepared PANI paper strip was first attached to a piece of stencil with the size of 1 cm × 3 cm using a double-sided tape. Copper tape were cut into pieces with the size of 1 cm × 0.635 cm and then placed on both sides of the PANI paper strip on the stencil as the electrodes, leaving a 1-mm gap between the PANI paper strip and each copper electrode. The PANI paper strip and the copper electrodes were physically connected by the conductive silver ink. The whole device was then kept at RT for at least 12 h to allow the silver ink to dry and ensure good conductivity between the paper strip and the electrodes. The NIP-PANI paper sensors were fabricated following the same procedures above as control.

2.4. Measurement of MMA Concentration

Aqueous solutions with various MMA concentrations were prepared with DI water. The stock solution was prepared by dissolving 200 µM MMA in human plasma solution. Sample solutions with a series of concentrations of 0, 20, 40, 60, 80, and 100 µM were prepared by diluting the MMA stock solution by human plasma. Then, 10 µL of these sample solutions were dispensed on the surface of the MMA-MIP-PANI paper sensors. The direct current resistance (R) of the paper sensor was measured by a multimeter (8846A, Fluke, Everett, WA, USA) before and 30 min after sample dispensing. The resistance change of the NIP-PANI paper sensor was also recorded for each MMA concentration as control.

The MMA concentration is determined by the resistivity (ρ) change of the paper sensor before and after exposure to the MMA sample solution. As we know, the ρ of the paper sensor is related to R and can be described by Equation (1),

$$\rho = R \cdot \frac{A}{l} \tag{1}$$

where A is the cross-sectional area and l is the length of the PANI paper strip between the electrodes. Since A and l remain the same for each paper sensor before and after sample dispensing, the R is positively proportional to the ρ of the paper sensor. As a result, the resistivity change of the paper sensor is positively related to the resistance change of the paper sensor. The resistance change ratio (ΔR) after and before the sample exposure is normalized based on the MMA concentration at 0 mg/L by the following Equation (2).

$$\Delta R = \left(\frac{\rho_{after,sample}}{\rho_{before,sample}} \right) \Big/ \left(\frac{\rho_{after,DI\ water}}{\rho_{before,DI\ water}} \right) \tag{2}$$

2.5. Data Analysis

The limit of detection (LoD) of the MMA-MIP-PANI paper sensor is evaluated by the following method. First, the limit of blank (*LoB*) is determined by the equation [22],

$$LoB = \mu_b + \sigma_b \tag{3}$$

where μ_b and σ_b are the mean value and the standard deviation of blank samples, respectively. Then, the LoD is calculated by the equation [22],

$$LoD = LoB + \sigma_s \tag{4}$$

where σ_s is the standard deviation of the sample with a low concentration. By plugging in the mean value and standard deviation from the calibration curve into Equations (3) and (4), the LoD of the MMA-MIP-PANI paper sensor can be estimated.

Each paper sensor was dispensed with the sample solution and the impedance change of the sensor was recorded. Three repeating measurements ($n = 3$) on different paper sensors were conducted on each glucose concentration. Data analysis was conducted by the Origin software.

3. Results and Discussion

3.1. Resistance

Before measurement, the resistance of the as-prepared MMA-MIP-PANI paper sensors were evaluated. Figure 1 shows the measured resistance of each paper sensor normalized to the average value. This result shows that the resistance of the MMA-MIP-PANI paper sensor was stable with small variations.

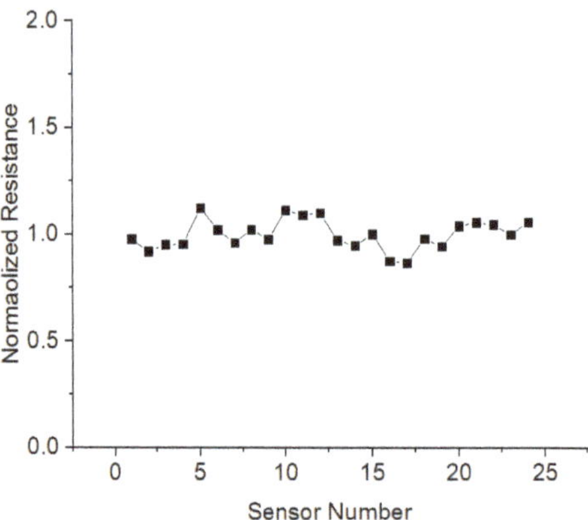

Figure 1. Normalized resistance distribution of the MMA-MIP-PANI paper sensors.

3.2. MMA Detection in Aqueous Solution

As seen in Figure 2, the resistance of the MMA-MIP-PANI paper sensor increases with the concentration of MMA, while the resistance of the NIP-PANI paper sensor kept almost constant. According to the linear fitting by the Origin software, a calibration curve as a function of MMA concentration in DI water was acquired with a correlation coefficient of 0.962. The resistance of the NIP-PANI paper sensor.

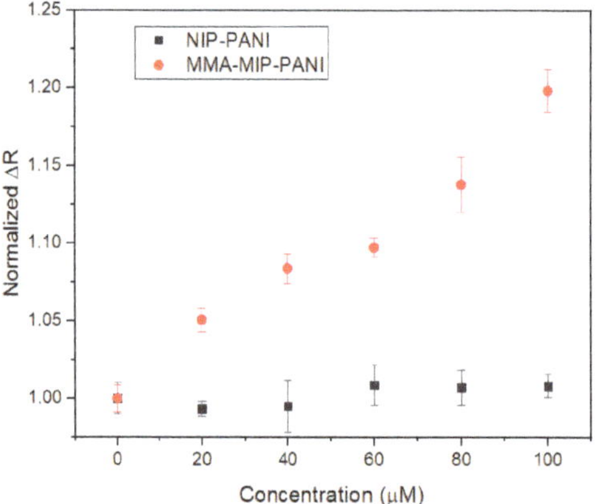

Figure 2. Normalized resistance ratio changes with the added MMA concentrations in DI water. Each point with error bar represents the average of three identical measurements with its standard deviation.

3.3. MMA Detection in Plasma

Figure 3 shows that the normalized resistance ratio of MMA-MIP-PANI paper sensor increases as the concentration of MMA increases. According to the linear fitting results, a calibration curve as a function of MMA concentration in plasma was obtained with a correlation coefficient of 0.987. The standard deviation of normalized resistance ratio at 0 µM was 0.0065. The mean normalized resistance ratio at 20 µM was 1.0453 with standard deviation of 0.0066. Thus, the LoD is also estimated to be 0.197 µM by the equations mentioned in the methods part. This result demonstrated the detection of the added MMA in human plasma with a wide linear range from 0 to 100 µM.

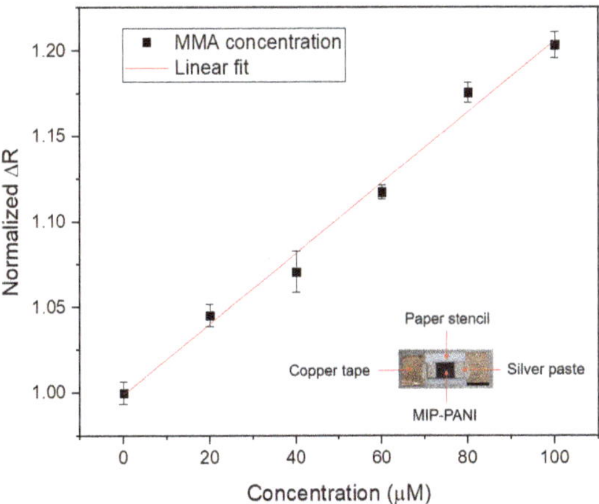

Figure 3. Normalized resistance ratio changes with the increased MMA concentration in plasma. Each point with error bar represents the average of three identical measurements with its standard deviation. Insert: a photograph of the MMA MIP-PANI paper sensor. Scale bar: 5 mm.

4. Conclusions

In summary, we proposed a simple and low-cost method for detecting MMA with a broad linear range by polyaniline paper sensor. The polyaniline paper strip was fabricated by a one-step solution process using MMA as the template by molecular imprinting technology. The concentration of MMA was determined by the resistance change of the paper sensor. A calibration curve as a function of MMA concentration in aqueous solution was acquired with a correlation coefficient of 0.962. We have demonstrated detection of the added MMA in plasma with a wide concentration range of 0 to 100 μM with a LoD of 0.197 μM. This disposable paper sensor is a promising alternative for MMA detection in cancer prognostics and diagnostics, especially for underserved communities.

Author Contributions: Conceptualization, Z.C. and J.K.; methodology, Z.C. and T.-Y.C.; formal analysis, Z.C.; investigation, Z.C. and T.-Y.C.; writing—original draft preparation, Z.C.; writing—review and editing, Z.C., T.-Y.C., and J.K.; supervision, J.K.; funding acquisition, J.K. All authors have read and agreed to the published version of the manuscript.

Funding: This research was funded by the Bill and Melinda Gates Foundation Grant OPP 1199456 and the National Science Foundation (NSF) Precise Advanced Technologies and Health Systems for Underserved Populations (PATHS-UP) Grant 1648451.

Data Availability Statement: The data presented in this study are available in the article.

Conflicts of Interest: The authors declare no conflict of interest.

References

1. Clarke, R.; Sherliker, P.; Hin, H.; Nexo, E.; Hvas, A.M.; Schneede, J.; Birks, J.; Ueland, P.M.; Emmens, K.; Scott, J.M.; et al. Detection of vitamin B12 deficiency in older people by measuring vitamin B12 or the active fraction of vitamin B12, holotranscobalamin. *Clin. Chem.* **2007**, *53*, 963–970. [CrossRef]
2. Hempen, C.; Wanschers, H.; Veer, G.V. A fast liquid chromatographic tandem mass spectrometric method for the simultaneous determination of total homocysteine and methylmalonic acid. *Anal. Bioanal. Chem.* **2008**, *391*, 263–270. [CrossRef]
3. Tecleab, A.G.; Schofield, R.C.; Ramanathan, L.V.; Carlow, D.C. A Simple and Sensitive Method for Quantitative Measurement of Methylmalonic Acid by Turbulent Flow Chromatography and Tandem Mass Spectrometry. *J. Chromatogr. Sep. Tech.* **2016**, *7*, 336. [CrossRef] [PubMed]
4. Gomes, A.P.; Ilter, D.; Low, V.; Endress, J.E.; Fernández-García, J.; Rosenzweig, A.; Schild, T.; Broekaert, D.; Ahmed, A.; Planque, M.; et al. Age-induced accumulation of methylmalonic acid promotes tumour progression. *Nature* **2020**, *585*, 283–287. (In English) [CrossRef] [PubMed]

5. Schroder, T.H.; Quay, T.A.W.; Lamers, Y. Methylmalonic Acid Quantified in Dried Blood Spots Provides a Precise, Valid, and Stable Measure of Functional Vitamin B-12 Status in Healthy Women. *J. Nutr.* **2014**, *144*, 1658–1663. (In English) [CrossRef]
6. Obeid, R.; Geisel, J.; Herrmann, W. Comparison of two methods for measuring methylmalonic acid as a marker for vitamin B12 deficiency. *Diagnosis* **2015**, *2*, 67–72. (In English) [CrossRef]
7. Suhito, I.R.; Koo, K.M.; Kim, T.H. Recent Advances in Electrochemical Sensors for the Detection of Biomolecules and Whole Cells. *Biomedicines* **2020**, *9*, 15. [CrossRef] [PubMed]
8. Tiwari, J.N.; Vij, V.; Kemp, K.C.; Kim, K.S. Engineered Carbon-Nanomaterial-Based Electrochemical Sensors for Biomolecules. *ACS Nano* **2016**, *10*, 46–80. [CrossRef] [PubMed]
9. Rajkumar, C.; Veerakumar, P.; Chen, S.M.; Thirumalraj, B.; Liu, S.B. Facile and novel synthesis of palladium nanoparticles supported on a carbon aerogel for ultrasensitive electrochemical sensing of biomolecules. *Nanoscale* **2017**, *9*, 6486–6496. [CrossRef] [PubMed]
10. Erturk, G.; Mattiasson, B. Molecular Imprinting Techniques Used for the Preparation of Biosensors. *Sensors* **2017**, *17*, 288. [CrossRef]
11. Ho, K.-C.; Yeh, W.-M.; Tung, T.-S.; Liao, J.-Y. Amperometric detection of morphine based on poly(3,4-ethylenedioxythiophene) immobilized molecularly imprinted polymer particles prepared by precipitation polymerization. *Anal. Chim. Acta* **2005**, *542*, 90–96. [CrossRef]
12. BelBruno, J.J. Molecularly Imprinted Polymers. *Chem. Rev.* **2019**, *119*, 94–119. [CrossRef]
13. Hua, D.; Lin, L.; Mizhi, J.; Shuangmei, Z.; Mingbo, Y.; Qiang, F. Progress on the morphological control of conductive network in conductive polymer composites and the use as electroactive multifunctional materials. *Prog. Polym. Sci.* **2014**, *39*, 627–655. [CrossRef]
14. Masoudi, F.M.; Leila, G.; Farideh, P.; Abbasali, Z.; Fariba, S. Molecular imprinting method for fabricating novel glucose sensor: Polyvinyl acetate electrode reinforced by MnO_2/CuO loaded on graphene oxide nanoparticles. *Food Chem.* **2016**, *194*, 61–67. [CrossRef]
15. Yoshimi, Y.; Narimatsu, A.; Nakayama, K.; Sekine, S.; Hattori, K.; Sakai, K. Development of an enzyme-free glucose sensor using the gate effect of a molecularly imprinted polymer. *J. Artif. Organs* **2009**, *12*, 264. [CrossRef]
16. Chen, Z.; Chi, T.Y.; Dincel, O.; Tong, L.; Kameoka, J. A Low-cost and Enzyme-free Glucose Paper Sensor. In Proceedings of the 2020 42nd Annual International Conference of the IEEE Engineering in Medicine & Biology Society (EMBC), Montreal, QC, Canada, 20–24 July 2020; pp. 4097–4100. [CrossRef]
17. Zhong, C.; Yang, B.; Jiang, X.; Li, J. Current Progress of Nanomaterials in Molecularly Imprinted Electrochemical Sensing. *Crit. Rev. Anal. Chem.* **2018**, *48*, 15–32. [CrossRef]
18. Akshaya, K.B.; Anitha, V.; Nidhin, M.; Sudhakar, Y.N.; Louis, G. Electrochemical sensing of vitamin B12 deficiency marker methylmalonic acid using PdAu-PPy tailored carbon fiber paper electrode. *Talanta* **2020**, *217*, 121028. [CrossRef]
19. Deepa, J.R.; Anirudhan, T.S.; Gowri, S.; Sekhar, V.C. Electrochemical sensing of methylmalonic acid based on molecularly imprinted polymer modified with graphene oxide and gold nanoparticles. *Microchem. J.* **2020**, *159*, 105489. [CrossRef]
20. Chen, Z.; Wright, C.; Dincel, O.; Chi, T.Y.; Kameoka, J. A Low-Cost Paper Glucose Sensor with Molecularly Imprinted Polyaniline Electrode. *Sensors* **2020**, *20*, 1098. [CrossRef] [PubMed]
21. Chi, T.Y.; Chen, Z.; Kameoka, J. Perfluorooctanesulfonic Acid Detection Using Molecularly Imprinted Polyaniline on a Paper Substrate. *Sensors* **2020**, *20*, 7301. [CrossRef]
22. Borysiak, M.D.; Thompson, M.J.; Posner, J.D. Translating diagnostic assays from the laboratory to the clinic: Analytical and clinical metrics for device development and evaluation. *Lab Chip* **2016**, *16*, 1293–1313. (In English) [CrossRef] [PubMed]

Proceeding Paper

A Complementary Reduced Graphene Oxide-Based Inverter for Ion Sensing [†]

Rassen Boukraa *, Giorgio Mattana, Nicolas Battaglini * and Benoit Piro

ITODYS, Université Paris Cité, CNRS, F-75006 Paris, France; giorgio.mattana@u-paris.fr (G.M.);
piro@u-paris.fr (B.P.)
* Correspondence: rassen.boukraa@gmail.com (R.B.); nicolas.battaglini@u-paris.fr (N.B.)
† Presented at the 2nd International Electronic Conference on Biosensors, 14–18 February 2022; Available online:
https://sciforum.net/event/IECB2022.

Abstract: Graphene, a 2D material with high conductivity and stability in aqueous media, could complement silicon as raw material for sensing with transistor-based devices in liquids. Furthermore, the fabrication of graphene-based transistors is affordable with low-cost techniques such as inkjet printing from graphene oxide (GO)-based inks. Deposited on plastic conformable substrates, graphene-based logic gates are standing as favorable and compelling candidates in the field of biosensing, to make electrical transduction and binary operations match with aqueous media and facilitate diagnostic operations.

Keywords: electrolyte-gated field-effect transistors; chemical logic; reduced graphene oxide-based inverters; sweat sensor

Citation: Boukraa, R.; Mattana, G.; Battaglini, N.; Piro, B. A Complementary Reduced Graphene Oxide-Based Inverter for Ion Sensing. *Eng. Proc.* **2022**, *16*, 2. https://doi.org/10.3390/IECB2022-12272

Academic Editors: Giovanna Marrazza and Sara Tombelli

Published: 14 February 2022

Publisher's Note: MDPI stays neutral with regard to jurisdictional claims in published maps and institutional affiliations.

Copyright: © 2022 by the authors. Licensee MDPI, Basel, Switzerland. This article is an open access article distributed under the terms and conditions of the Creative Commons Attribution (CC BY) license (https://creativecommons.org/licenses/by/4.0/).

1. Introduction

Graphene-based transistors are well known for being ambipolar devices stable in aqueous media, with high channel conductivity and charge carrier mobility, in comparison with organic semiconductor-based ones [1]. Electrolyte-gated graphene-based field-effect transistors (EG-GFETs) are composed of three electrodes, named source, drain, and gate, as in the case of conventional FETs, but they possess a few specificities of their own [2]. First of all, graphene cannot be considered as a semiconductor regarding its electrical behavior; in addition to that, the electrolyte separating the gate from the transistor's channel is easily tunable and modifiable, and that stands for both dielectric and sensing medium [3]. This latter property is particularly remarkable as the nature of the electrolyte induces a change in the electrochemical double-layer capacitance at the gate–electrolyte and graphene–electrolyte interfaces, which, in turn, leads to a modification of the electrical properties of the transistor itself [4]. The use of graphene transistors as biological or ionic sensors has already been highly reported [5]; however, logic gates built from graphene transistors have never been described for such purpose, even though few articles are already revealing research interest in such devices to perform basic logic operations in aqueous media [6]. In this study, we investigate the design of a logic gate for ion-sensing purposes from graphene-based electrical devices on flexible substrates, to further extend the application field to skin patches or bandage-type sensors, for instance.

2. Materials and Method

Transistors were designed with a coplanar gate configuration (Figure 1, left). The fabrication steps, from electrode patterning by photolithography to the electrochemical reduction of the GO ink-coated channel by drop-casting or inkjet printing method, are described elsewhere [7]. All electrolytes used were DI-water-based solutions, containing K^+ and/or Na^+ at various concentrations. Valinomycin (VMC) ion-selective membrane, specific to K^+, was deposited on the gate of each transistor by using the KELENN DMD100 printer. The composition was adapted from the method of Kisiel et al. [8] as follows: 1.0% (*w/w*) of valinomycin; 0.5% of potassium tetrakis [3,5-bis(trifluoromethyl) phenyl]borate (KTBP) selectophore to improve VMC selectivity; 31.5% of poly (vinyl chloride) polymer and 67.0% of 2-nitrophenyl octyl ether plasticizer in cyclohexanone instead of THF as a solvent to prevent fast evaporation in the dispenser syringe. The conventional silicon wafer playing the role of the substrate was replaced here by a flexible polymer foil of polyimide to match with on-skin patch application perspectives. A commercial silver ink (purchased from PV Nano Cell) allowed the interconnections between the transistors. All electrical measurements were performed with a 4200-A-SCS Keithley analyzer. The biased supply voltage to acquire inverter characteristics was applied through an external power source (Basetech BT-305 DC).

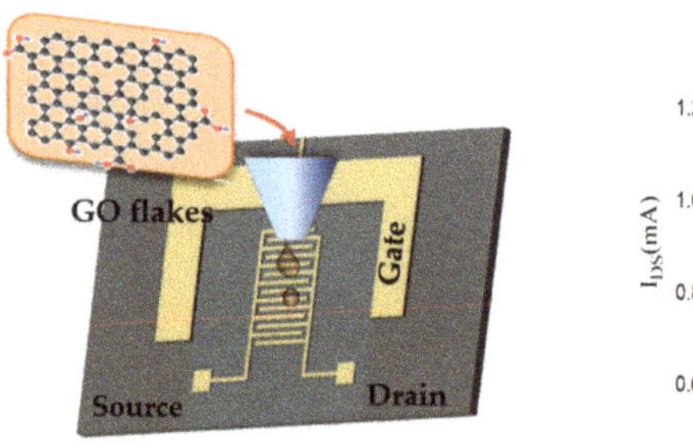

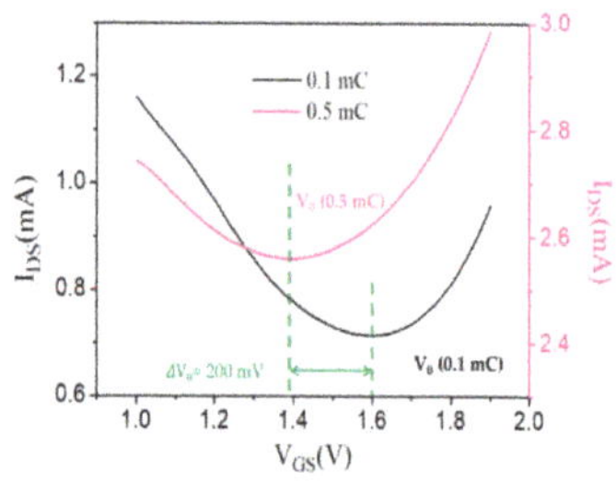

Figure 1. Coplanar gate configuration (**left**) and transfer characteristics for different charges during the electrochemical reduction process (**right**).

3. Results and Discussion

3.1. Electrical Characteristics of the EG-GFET Inverter

Regarding the ambipolar behavior of rGO, one expects an increase in drain current I_{DS}, apart from a minimum value that occurs at $V_{GS} = V_0$, corresponding to the situation where the Fermi level crosses the so-called Dirac point [1]. The electrochemical reduction of GO to rGO was controlled by measuring the amount of charge passed at a constant reduction voltage, showing good control of the $V_0 = V_{GS}$ (at $I_{DS,min}$). V_0 can be tuned as shown in Figure 1, right; the higher the amount of charge, the less positive the value of V_0. However, even with a significant value of charge injected, V_0 stays positive, which indicates that the rGO built following this process is still p-doped [5,7].

The difference in V_0 from one to another transistor may be caused by a difference in the doping state. Therefore, a slight shift of a few tens of mV is enough to build an inverter out of two graphene transistors made from two different reduction charges: The two gates, as well as the drains, were connected together so that the transistor with the highest V_0 acted as the *p* transistor and connected to V_{DD} from its source; the second one, still p-doped even if less, was connected to the ground and acted as the *n* transistor of a usual NOT gate. The electrical characteristic of the resulting gate is shown in Figure 2.

3.2. The EG-GFET Inverter as K$^+$ Sensor

One way to study inverters is to consider each of the transistors T1 and T2 as a variable resistor. The competition between those two connected transistors is related to the ratio R_{T1} versus R_{T2}: when the upper transistor, connected to the biased supply voltage V_{DD}, is driving the device, i.e., $R_{T1} < R_{T2}$ occurring for low input V_{IN}, the output is connected to the supply voltage corresponding to the ON state. On the other hand, when $R_{T2} < R_{T1}$, the bottom transistor connects the ground to the output resulting in the OFF state.

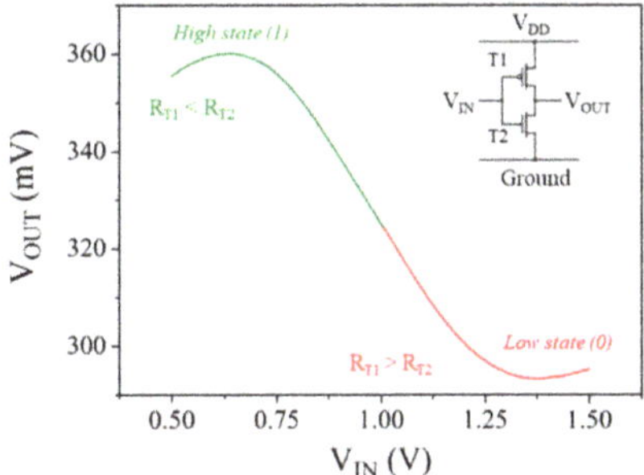

Figure 2. Graphene-based inverter input (IN) versus output (OUT) characteristic. Inset: electrical circuit with T1 (p-type, slightly reduced) and T2 transistors ("n-type-like", more reduced; see text for explanation).

It is clearly evident from above that different reduction levels lead to a shift in the V_0 of the two transistors; however, this shift can also be obtained by changing the capacitance at the gate–electrolyte interface, which is the way that we investigated in order to build a proper sensor out of the rGO inverter.

Functionalization of the gate electrode with ionophores or other specific hydrogel modifies the capacitance at the gate–electrolyte interface that induces a change in the effective gate voltage at the input of the modified transistor.

Figure 3 gives an example of ion sensing with rGO transistors, as the transfer characteristic is directly impacted by a change in the concentration of K^+ ions of the electrolyte, from 10^{-6} to 10^{-4} M, with a VMC gate-modified transistor.

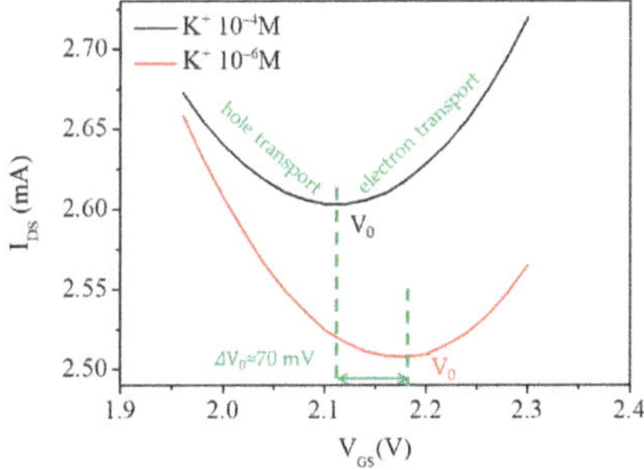

Figure 3. Transfer characteristic of VMC-modified rGO-EGFET for two different concentrations of K^+ ion.

We plan to use the rGO-based inverters to assist ion sensing, following the above-mentioned principle, which consists of performing the switch from OFF state to ON state, for a fixed V_{IN}, owing to the capacitance changes occurring at the gate–electrolyte interface of one of the two transistors, resulting in a change in the V_0 value. In this configuration, the ion concentration involved in the capacitance change is standing for the input information of the sensor. The main interest of sensing with inverters instead of isolated transistors is to take advantage of the Boolean algebra applied to graphene logic gates, which could allow us to design complex circuits stable in aqueous media and thus to perform chemical logic, which is an ingenious method to produce basic mathematical operations in solution, or, more concretely, to obtain a unique sensor built from the output responses of several ones in the same medium.

4. Conclusions

In this study, we reported a way to perform ion sensing with the most elementary logic gate—the inverter—from two interconnected transistors. To overcome heavy and costly processes of graphene deposition, we formulated a graphene oxide surfactant-free ink that can be directly inkjet printed on the channel. As an ongoing study, we are currently investigating the way to make the inverter switch occur, thereby estimating the sufficient threshold value of capacitance or ion concentration change needed to change from OFF to ON output state, through both experimental work and simulation.

Author Contributions: Conceptualization, G.M., N.B. and B.P.; methodology, R.B., G.M., N.B. and B.P.; investigation, R.B.; writing, R.B., G.M., N.B. and B.P.; supervision, G.M., N.B. and B.P. All authors have read and agreed to the published version of the manuscript.

Funding: This research received funding from the Consortium of French higher education and research institutions for the development of USTH, under the framework "Collaborative S&T projects 2021", USTH.AMSN.01/21-23, entitled "Functionalized Graphene for Electrolyte-Gated Field-Effect Transistors. Application to Wearable Sweat Sensors".

Institutional Review Board Statement: Not applicable.

Informed Consent Statement: Not applicable.

Acknowledgments: R.B thanks Ecole Doctorale ED388 *Chimie Physique et Chimie Analytique de Paris Centre* for providing a PhD grant.

Conflicts of Interest: The authors declare no conflict of interest.

References

1. Hess, L.H.; Seifert, M.; Garrido, J.A. Graphene Transistors for Bioelectronics. *Proc. IEEE* **2013**, *101*, 1780–1792. [CrossRef]
2. Ohno, Y.; Maehashi, K.; Yamashiro, Y.; Matsumoto, K. Electrolyte-Gated Graphene Field-Effect Transistors for Detecting pH and Protein Adsorption. *Nano Lett.* **2009**, *9*, 3318–3322. [CrossRef] [PubMed]
3. Kireev, D.; Brambach, M.; Seyock, S.; Maybeck, V.; Fu, W.; Wolfrum, B.; Offenhäusser, A. Graphene transistors for interfacing with cells: Towards a deeper understanding of liquid gating and sensitivity. *Sci. Rep.* **2017**, *7*, 6658. [CrossRef] [PubMed]
4. Heller, I.; Chatoor, S.; Mannik, J.; Zevenbergen, M.A.; Dekker, C.; Lemay, S.G. Influence of Electrolyte Composition on Liquid-Gated Carbon Nanotube and Graphene Transistors. *J. Am. Chem. Soc.* **2010**, *132*, 17149–17156. [CrossRef] [PubMed]
5. Furlan de Oliveira, R.; Livio, P.A.; Montes-García, V.; Ippolito, S.; Eredia, M.; Fanjul-Bolado, P.; Gonzalez Garcia, M.B.; Casalini, S.; Samorì, P. Liquid-Gated Transistors Based on Reduced Graphene Oxide for Flexible and Wearable Electronics. *Adv. Funct. Mater.* **2019**, *29*, 1905375. [CrossRef]
6. Son, M.; Kim, H.; Jang, J.; Kim, S.Y.; Ki, H.C.; Lee, B.H.; Kim, I.S.; Ham, M.H. Low-Power Complementary Logic Circuit Using Polymer-Electrolyte-Gated Graphene Switching Devices. *ACS Appl. Mater. Interfaces* **2019**, *11*, 47247–47252. [CrossRef] [PubMed]
7. Vasilijević, S.; Mattana, G.; Anquetin, G.; Battaglini, N.; Piro, B. Electrochemical tuning of reduced graphene oxide in printed electrolyte-gated transistors. Impact on charge transport properties. *Electrochim. Acta* **2021**, *371*, 137819. [CrossRef]
8. Kisiel, A.; Kałuża, D.; Paterczyk, B.; Maksymiuk, K.; Michalska, A. Quantifying plasticizer leakage from ion-selective membranes—A nanosponge approach. *Analyst* **2020**, *145*, 2966–2974. [CrossRef] [PubMed]

Proceeding Paper

Highly Sensitive Amperometric Biosensors Based on Oxidases and CuCe Nanoparticles Coupled with Porous Gold [†]

Nataliya Stasyuk [1,2], Olha Demkiv [1,3], Galina Gayda [1,*], Roman Serkiz [1], Andriy Zakalskiy [1], Oksana Zakalska [1], Halyna Klepach [2], Galeb Al-Maali [4], Nina Bisko [4] and Mykhailo Gonchar [1,2]

[1] Institute of Cell Biology National Academy of Sciences of Ukraine, 79005 Lviv, Ukraine; stasukne@nas.gov.ua (N.S.); demkivo@nas.gov.ua (O.D.); rserkiz@gmail.com (R.S.); zakalsky@yahoo.com (A.Z.); zakalska@yahoo.com (O.Z.); gonchar@cellbiol.ua (M.G.)

[2] Department of Biology and Chemistry, Drohobych Ivan Franko State Pedagogical University, 82100 Drohobych, Ukraine; pavlishko@yahoo.com

[3] Faculty of Veterinary Hygiene, Ecology and Law, Stepan Gzhytskyi National University of Veterinary Medicine and Biotechnologies, 79000 Lviv, Ukraine

[4] M. G. Kholodny Botany Institute, National Academy of Sciences of Ukraine, 01601 Kyiv, Ukraine; galmaali@nas.gov.ua (G.A.-M.); bisko_nina@ukr.net (N.B.)

* Correspondence: galina.gayda@gmail.com; Tel.: +380-322-612-144

† Presented at the 2nd International Electronic Conference on Biosensors, 14–18 February 2022; Available online: https://sciforum.net/event/IECB2022.

Abstract: Metallic nanoparticles are usually applied in biosensors as catalysts and/or mediators of electron transfer. We describe the development of amperometric biosensors (ABSs) based on oxidases and nanoparticles of CuCe (nCuCe). nCuCe, being an electro-active mediator and active peroxidase (PO) mimetic, was used as an H_2O_2-sensing platform in oxidase-based ABSs. ABSs for glucose, primary alcohols, methyl amine, catechol, and L-arginine, which are based on corresponding oxidases and nCuCe, were developed. These ABSs exhibited improved analytical characteristics in comparison with the appropriate bi-enzyme ABSs containing natural PO. Including electrode-posited porous gold in the chemo-sensing layer was shown to increase significantly the sensitivities of all constructed ABSs.

Keywords: electroactive nanoparticles; peroxidase-like nanozyme; oxidases; porous gold; amperometric biosensors

Citation: Stasyuk, N.; Demkiv, O.; Gayda, G.; Serkiz, R.; Zakalskiy, A.; Zakalska, O.; Klepach, H.; Al-Maali, G.; Bisko, N.; Gonchar, M. Highly Sensitive Amperometric Biosensors Based on Oxidases and CuCe Nanoparticles Coupled with Porous Gold. *Eng. Proc.* **2022**, *16*, 3. https://doi.org/10.3390/IECB2022-12251

Academic Editors: Giovanna Marrazza and Sara Tombelli

Published: 14 February 2022

Publisher's Note: MDPI stays neutral with regard to jurisdictional claims in published maps and institutional affiliations.

Copyright: © 2022 by the authors. Licensee MDPI, Basel, Switzerland. This article is an open access article distributed under the terms and conditions of the Creative Commons Attribution (CC BY) license (https://creativecommons.org/licenses/by/4.0/).

1. Introduction

Metallic nanoparticles have wide potential practical applications in various fields of science and industry. In biosensorics, they usually act as mediators in electron transfer and/or catalysts (nanozymes, NZ) [1–5].

NZs are the newest class of functional nanomaterials [3–7] that have enzyme-like activities with different reaction specificities. NZs possess increased stability and greater availability due to their simpler preparation technologies. Most reported NZs are mainly mimetics of oxidoreductases, including peroxidase (PO) [7–9].

PO catalyzes the oxidation of diverse organic compounds, using H_2O_2 as the electron acceptor [8]. Many natural enzymes (oxidases) produce H_2O_2 as a byproduct of their enzymatic reactions, so the detection of a target substrate can be performed by measuring H_2O_2 generation. Over the last few years, a number of reports have described the application of various PO-like NZs for H_2O_2 detection using different sensors [10–14]. The main peculiarities of PO-like NZs as catalysts are that they have high stability, sensitivity, and selectivity to H_2O_2 in extra-wide linear ranges. PO-like NZs coupled with natural oxidases are widely used in amperometric oxidase-based biosensors (ABSs) [1–3,7–10].

In our earlier works, different types of chemically and "green" synthesized PO-like NZs were described [9,10]. Nanoparticles of CuCe (nCuCe) were chosen as effective electroactive PO-mimetics and were characterized using scanning electron microscopy (SEM) coupled with X-ray microanalysis (SEM-XRM) [9,15]. Our results demonstrated that the synthesized nCuCe, having an excellent sensitivity and a wide linear range for H_2O_2 detection, may be promising artificial POs for the development of oxidase-based ABSs. nCuCe were successfully used for the construction of Arg-sensitive ArgO-based ABSs [15].

A number of approaches have been proposed for improving the analytical characteristics of ABSs. One of them is increasing the effective working surface of the electrode in order to obtain maximal electroactive sites for the immobilization of biocatalysts, including enzymes and NZs [7,8].

Micro/nanoporous gold (npAu), because of its high surface area-to-volume ratio, excellent conductivity, chemical inertness, physical and chemical stability, biocompatibility, high area, electrochemical activity, easily tunable pores, and plasmonic properties, may be promising in medicine for diagnostics and drug delivery, in energy storage, and in sensing and biosensing. A lot of synthetic methods for obtaining npAu have been reported, including deallying, templating, sputtering, self-assembling, and electrodeposition [16,17]. The last method is the most popular.

The aim of the current study was to fabricate and characterize highly sensitive ABSs using various oxidases as biorecognition elements with nCuCe as an electroactive mimetic of PO and electrodeposited npAu as an effective carrier of enzymes/NZs with a highly advanced surface area [16,17].

2. Materials and Methods

2.1. Reagents

Cerium(IV) bicarbonate($Ce(HCO_3)_4$), copper(II) sulfate ($CuSO_4$), L-arginine (Arg), methylamine (MA), ethanol, methanol, *o*-dianisidine, hydrogen peroxide (H_2O_2, 30%), hydrogen tetrachloroaurumate(III) $H[AuCl_4]$, D-glucose, sodium sulfide (Na_2S), ammonia chloride (NH_4Cl), Nafion (5% solution in 90% low-chain aliphatic alcohols), Horse radish peroxidase (PO, EC 1.11.1.7) from *Armoracia rusticana* (500 $U \cdot g^{-1}$), glucose oxidase (GO, EC 1.1.3.4) from *Aspergillus niger* (168 $U \cdot mg^{-1}$), and all other reagents and solvents used in this work were purchased from Sigma-Aldrich (Steinheim, Germany).

All reagents were of analytical grade and were used without further purification. All solutions were prepared using ultra-pure water obtained with the Milli-Q® IQ 7000 Water Purification system (Merck KGaA, Darmstadt, Germany).

2.2. Enzymes Isolation and Purification

Electrophoretically homogeneous yeast enzymes—alcohol oxidase (AO, EC1.1.3.13), L-arginine oxidase (ArgO, EC 1.4.3.25), methylamine oxidase (AMO, EC 1.4.3.21), and laccase (EC 1.10.3.2) were used for amperometric biosensors fabrication.

Yeast AO was isolated from cell-free extract of the selected over-producing strain *Ogataea polymorpha* C-105 (*gcr1 catX*) using a two-step ammonium sulfate fractionation (at 30 and 70% saturation), followed by ion exchange chromatography on DEAE-Toyopearl 650 M [18]. Purified AO with specific activity ~20 $U \cdot mg^{-1}$ of protein was kept as a suspension in 70% sulfate ammonium and 50 mM phosphate buffer (PB) at pH 7.5 at 4 °C.

Fungal ArgO was isolated from an extract of the fruiting body of the wild forest mushroom *Amanita phalloides* and partially purified up to ~7.9 $U \cdot g^{-1}$ of protein using a two-step ammonium sulfate fractionation (at double 70% of saturation), followed by ion exchange chromatography on Toyopearl DEAE-650M resin [15].

Activities of AO, ArgO, or GO were determined by the rate of hydrogen peroxide formation in reaction with substrate (methanol, Arg, or glucose) as monitored by the peroxidative oxidation of *o*-dianisidine in the presence of PO and correspondent substrates methanol [18], Arg [15], or glucose [10].

Yeast AMO was isolated from the recombinant yeast strain *Saccharomyces cerevisiae* C13ABYS86 [19]. The (His)$_6$-tagged AMO was purified from the cell-free extract by metal-affinity chromatography on Ni-NTA-agarose. Activity of AMO was determined by the rate of hydrogen peroxide formation in reaction with MA as monitored by the peroxidative oxidation of 2,2′-Azino-bis(3-ethylbenzthiazoline-6-sulfonic acid (ABTS) in the presence of PO.

Fungal laccase was isolated from a cultural liquid of the fungus *Trametes zonatus* by a two-step ammonium sulfate fractionation (up to 70% of saturation), followed by ion exchange chromatography on Toyopearl DEAE-650M [20]. Fractions with the laccase activity were pooled, concentrated by Millipore filter (10 kDa) up to specific activity of enzyme $\geq 10\ \mathrm{U \cdot mg^{-1}}$ followed by precipitation with 80% sulfate ammonium.

The activity of laccase was determined by the rate of the increase in absorbance monitored spectrophotometrically at 420 nm (Shimadzu, Kyoto, Japan). As a substrate, 0.5 mM ABTS in 50 mM sodium acetate (NaOAc) buffer solution, pH 4.5 was used. One unit of laccase activity was defined as the amount of the enzyme required to oxidize 1 µmole of a substrate ($\varepsilon_{420} = 36\ \mathrm{mM^{-1} \cdot cm^{-1}}$) per minute at 24 °C.

2.3. Synthesis of Nanoparticles and Estimation of Their Pseudo-Peroxidase Activity

Nanoparticles of CuCe (nCuCe) were synthesized, as described previously [9]. The synthesized nCuCe were collected by centrifugation. The precipitates were rinsed twice with water and were stored as a water suspension at +4 °C until use.

Pseudo-peroxidase (PO-like) activity of the nCuCe was measured using the colorimetric method, with *o*-dianisidine as a chromogenic substrate in the presence of H_2O_2 [9]. One unit (U) of PO-like activity was defined as the amount of nCuCe releasing 1 µmol H_2O_2 per 1 min at 30 °C under standard assay conditions.

2.4. Apparatus

A piece of Pt wire and an Ag/AgCI/3M KCI electrode were used as the counter and reference electrodes. 3.05 mm graphite rods (type RW001, Ringsdorff Werke, Bonn, Germany) were used as working electrodes. They were sealed in glass tubes with epoxy forming disk electrodes. Before sensor preparation, the graphite electrode (GE) was polished with emery paper. Amperometric measurements were carried out with a potentiostat CHI 1200A (IJ Cambria Scientific, Burry Port, UK) in batch mode under continuous stirring in a standard 40 mL cell at a room temperature.

A SEM microanalyser REMMA-102-02 (Lviv, Ukraine) was used for morphological analyses of the synthesized nAu-film (nAu).

2.5. Electrodeposition of Nanoporous Gold onto Graphite Electrode

A micro/nanoporous gold (npAu) was synthesized on the surface of GE in two stages. In the first stage, nAu was electrodeposited from a solution containing 10 mM of $HAuCl_4$ in 2.5 M ammonia chloride, using cyclic voltammetry in the range of 0 to +800 mV with a scan rate of $50\ \mathrm{mV \cdot min^{-1}}$ for 25 cycles. In the second stage, the obtained modified electrode (npAu/GE) was re-immersed in a solution of 10 mM of $HAuCl_4$ in 2.5 M of ammonia chloride using the potentiostatic mode at -1000 mV for 120 s. The obtained npAu/GE was rinsed with water and equilibrated before usage in the appropriate buffer.

2.6. Immobilization of Natural and Artificial Peroxidases onto Electrode

Natural PO and the synthesized nCuCe as an artificial PO were immobilized on the surfaces of GE or npAu/GE using the physical adsorption method. For the development of the nCuCe/npAu/GE, an aliquot of nCuCe solution (5–10 µL) with a PO-like activity of 1 U/mL was dropped onto the surface of npAu/GE. For development of the PO/GE, an aliquot of PO solution (5–10 µL) with an activity of 1 U/mL was dropped onto the surface of GE. After drying the sensing film for 10 min at room temperature, the modified GE was covered with 10 µL of 1% Nafion solution in 50 mM PB, pH7.5. The modified electrodes were rinsed with 50 mM PB, pH 7.5, and kept in this buffer with 0.1 mM EDTA at 4 °C until used.

2.7. Immobilization of Oxidases onto the Modified Electrodes

To fabricate the oxidase-based amperometric biosensors (ABS), GO, AMO, AO, ArgO, or laccase were immobilized onto the modified GE.

A total of 5–10 µL of enzyme solution was dropped onto the dried surfaces of the PO/GE, nCuCe/GE, or nCuCe/npAu/GE. To develop the ABSs on the base of ArgO, GO, MAO or laccase, the dried composites were covered with a Nafion membrane, as described in Section 2.6. To prepare 1% Nafion solution, the stock 5% solution was diluted with the appropriate buffer: 50 mM NaOAc, pH4.5 for construction of laccase-based ABS and with 50 mM PB, pH 7.5 in other cases.

It is worth mentioning that, in the AO-based ABS, the biosensing film on the electrode was fixed, not with Nafion, but with a dialysis membrane.

The coated bioelectrodes were rinsed with water and stored in the corresponding buffers until use.

2.8. Measurements and Calculations

Amperometric measurements were carried out using a CHI 1200A potentiostat (IJ Cambria Scientific, Burry Port, UK) connected to a personal computer, used in a batch mode under continuous stirring in an electrochemical cell with a 20 mL volume at 25 °C.

All experiments were carried out in triplicate trials. Analytical characteristics of the proposed electrodes were statistically processed using OriginPro 8.5 software. Error bars represent the standard error derived from three independent measurements. Calculation of the apparent Michaelis–Menten constants ($K_M{}^{app}$) was performed automatically by this program, according to the Lineweaver–Burk equation.

3. Results and Discussion

3.1. Development of Oxidase-Based Biosensors Using nCuCe and Porous Gold

We describe here the development of amperometric biosensors (ABSs) based on oxidases and nCuCe. nCuCe, being an active PO mimetic, was used as a hydrogen peroxide-sensing platform for oxidase-based ABSs.

To improve analytical characteristics of ABSs, namely, sensitivity, we modified the surface of a graphite electrode (GE) with micro/nanoporous gold (npAu).

npAu has a high surface area-to-volume ratio, excellent conductivity, chemical stability, high area, electrochemical activity, easily tunable pores, and plasmonic properties, thus, it may be promising for use in sensing and biosensing.

npAu has the unique properties of chemical stability, a high area, and electrochemical activity, thus it may be promising in biosensing. The principal scheme of bioelectrode construction is presented in Figure 1.

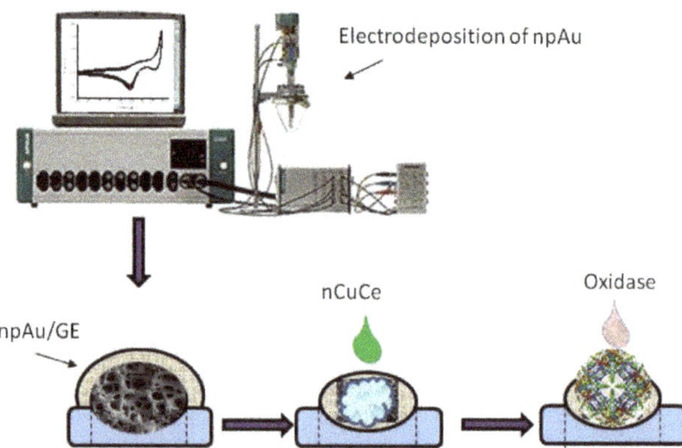

Figure 1. Scheme of electrode modification.

Figure 2 demonstrates the results of morphological characterization of npAu using the SEM technique, which provides information on the size, distribution, and shape of the tested npAu. The XRM images showed the characteristic peaks for the gold.

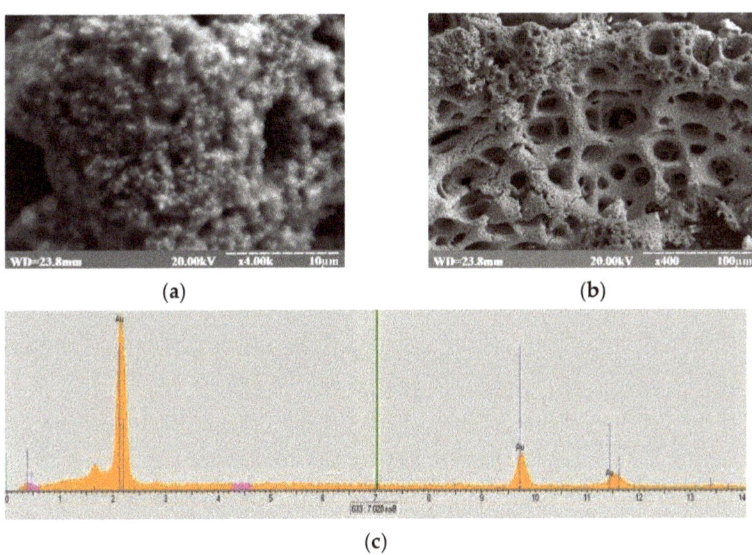

Figure 2. Characteristics of the npAu SEM images (**a**,**b**); X-ray spectral microanalysis (**c**).

3.2. Analytical Characteristics of the Constructed Biosensors

Using GO, AO, AMO, or ArgO as biorecognition elements, nCuCe as a PO-like NZ or as an electro-active mediator, and npAu as a carrier of enzymes/NZs, the ABSs for glucose, primary alcohols, methyl amine, L-arginine, and catechol, respectively, were constructed and characterized.

Figure 3 demonstrates the amperometric characteristics of the developed GO-based ABS for glucose determination. Using the chronoamperograms at optimal working potentials for the modified and control electrodes, calibration curves were plotted for analyte determination using the developed ABSs. The same experiments were carried out with other oxidase-based ABSs (data not shown). It is worth mentioning that the npAu/GE as a control electrode was also tested, and no amperometric signals were detected with any analyte addition under the chosen conditions (data not shown).

Table 1 summarizes the main bioanalytical characteristics for the developed ABSs, which were based on the usage of various oxidases and nanomaterials. It is worth mentioning that nCuCe plays a dual role in the developed ABSs: for laccase it is a mediator of electron transfer, and for other oxidases it is an artificial PO.

As can be seen in Table 1, nCuCe had a significant positive effect on sensor sensitivity in comparison to electrodes that were not modified with nanomaterials. For example, for AMO/nCuCe/GE and ArgO/nCuCe/GE, the sensitivities were 5-fold higher, than for the corresponding GEs without nCuCe.

The presence of npAu was shown to provide additional contributions to improving the analytical parameters of the ABS, especially in terms of their sensitivities. For example, the sensitivity of the GO/nCuCe/npAu/GE is 9.1-fold higher than that of the GO/PO/GE and 5.5-fold higher in comparison to the GO/nCuCe/GE. The same tendency, but at various levels, was demonstrated for all investigated enzymes. This fact has simple explanation: a highly advanced surface of the npAu, having hierarchical pores of nano- and micro-sizes with different diameters, has an enhanced working 3D surface area of electrode. The increased surface of the modified GE leads to the enhanced adsorption of nanomaterials/enzymes and, thus, to improved efficiency of electron transfer in ABS in comparison with unmodified GEs.

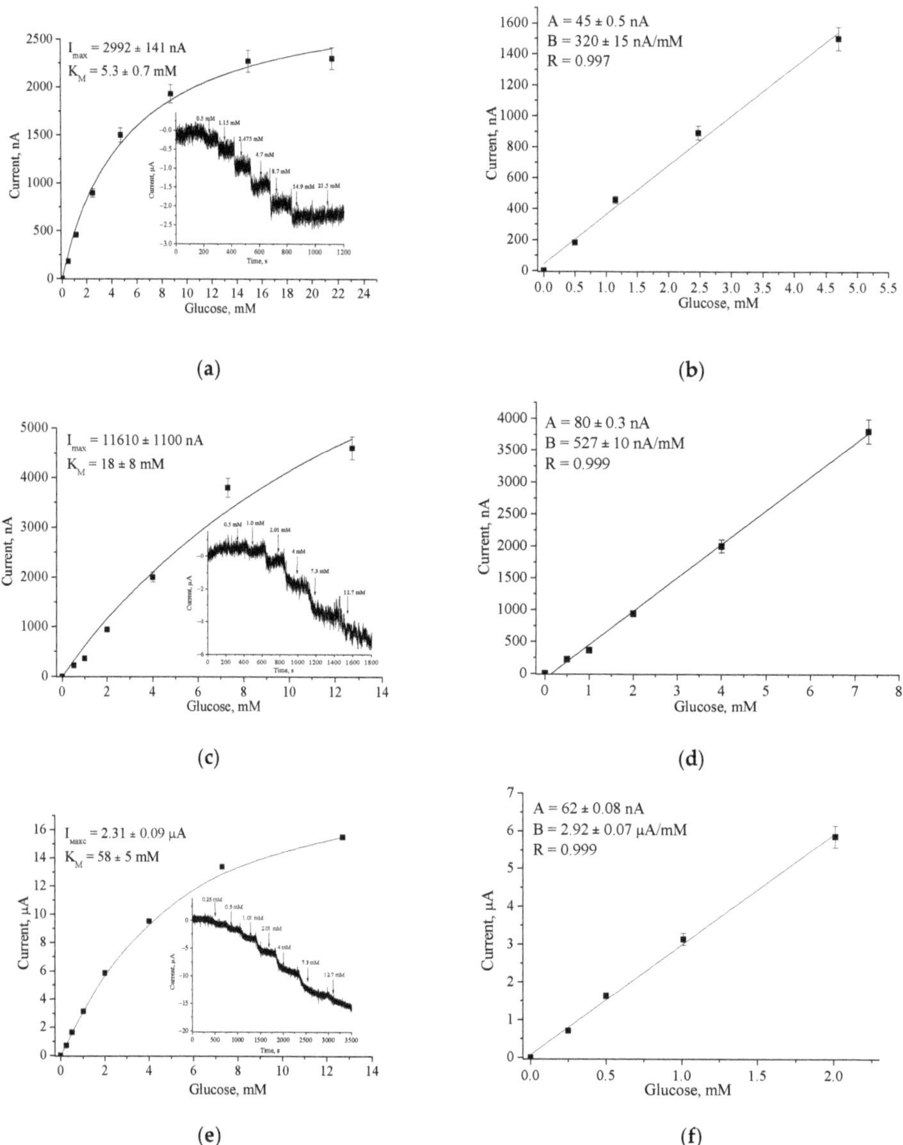

Figure 3. Amperometric characteristics of the GO/PO/GE (**a,b**), GO/nCuCe/GE (**c,d**) and GO/nCuCe/npAu/GE (**e,f**): (**a,c,e**) chronoamperogramms (inserted) and dependences of amperometric signal on concentration of glucose; (**b,d,f**) calibration graphs for glucose determination. Conditions: working potential –50 mV vs. Ag/AgCl/3 M KCl in 50 mM PB, pH 6.0. The sensing layers contain 0.01 U of PO/PO-like activity and 0.01 U of GO.

Table 1. Analytical characteristics of the constructed bioelectrodes based on different oxidases, natural or artificial peroxidases, and npAu.

Bioelectrode	Potential, mV	Sensitivity, $A \cdot M^{-1} \cdot m^{-2}$	Linear Range, μM	LOD, μM
GO/PO/GE	−50	44	50–5000	150
GO/nCuCe/GE	−50	73	500–7300	150
GO/nCuCe/npAu/GE	−50	400	25–2000	75.7
AMO/PO/GE	−250	7	200–1700	130
AMO/nCuCe/GE	−250	35	60–1700	18
AMO/nCuCe/npAu/GE	−250	125	60–500	18
AO/PO/GE	−50	22	130–900	39
AO/nCuCe/GE	−50	32	50–2100	15
AO/nCuCe/npAu/GE	−50	102	33–500	10
ArgO/PO/GE	−150	24	75–1150	35
ArgO/nCuCe/GE	−150	113	50–2250	15
ArgO/nCeCu/npAu/GE	−150	200	100–500	33
Laccase/GE	+200	2300	8–160	2
Laccase/nCuCe/GE	+200	5055	3–40	1.5
Laccase/nCuCe/npAu/GE	+200	9280	2–40	1

4. Conclusions

In the present work, the development of ABSs based on different oxidases and nCuCe was described. nCuCe has a dual role being an active mimetic of PO and a mediator of electron transfer. It was used as an electro-active mediator for laccase-based ABS and as a PO-like NZ in ABSs, based on other oxidases, namely, GO, AO, AMO and ArgO. The ABSs for catechol, glucose, primary alcohols, methyl amine, and L-arginine were constructed and characterized. The developed mono-enzyme ABSs exhibited improved analytical characteristics in comparison with the correspondent bi-enzyme ABSs, which contained natural PO. It was demonstrated, that including electrodeposited nanoporous gold in chemosensing layer on graphite electrode allows a significant additional increase in ABSs sensitivity. This fact may be explained by the highly advanced surface area of npAu due to pores of nano- and micro-sizes. Such a hierarchical porous 3D surface leads to the enhanced adsorption of nanomaterials/enzymes and, thus, to improved efficiency of electron transfer in ABS, in comparison with unmodified GEs.

Author Contributions: Conceptualization, N.S. and G.G.; methodology, N.S., O.D. and G.A.-M.; software, R.S.; validation, O.D. and H.K.; formal analysis, N.B.; investigation, O.D., O.Z., R.S., resources, O.Z.; data curation, A.Z. and N.B.; writing—original draft preparation, N.S.; writing—review and editing, G.G.; visualization, A.Z.; supervision, M.G.; project administration, M.G.; funding acquisition, M.G. All authors have read and agreed to the published version of the manuscript.

Funding: This research was partially founded by the National Research Foundation of Ukraine (projects No 2020.02/0100 and No 2021.01/0010), by the National Academy of Sciences of Ukraine (The program "Smart sensor devices of a new generation based on modern materials and technologies"), and Ministry of Education and Science of Ukraine (projects Nos. 0120U103398, 0121U109539 and 0121U109543).

Institutional Review Board Statement: Not applicable.

Informed Consent Statement: Not applicable.

Data Availability Statement: Data is contained within this article.

Acknowledgments: The authors thank M. Horecha (Institute of Cell Biology NAS of Ukraine, Lviv, Ukraine) for her technical assistance in experiments.

Conflicts of Interest: The authors declare no conflict of interest.

References

1. Naresh, V.; Lee, N. A review on biosensors and recent development of nanostructured materials-enabled biosensors. *Sensors* **2021**, *21*, 1109. [CrossRef] [PubMed]
2. Nayl, A.A.; Abd-Elhamid, A.I.; El-Moghazy, A.Y.; Hussin, M.; Abu-Saied, M.A.; El-Shanshory, A.A.; Solman, H.M.A. The nanomaterials and recent progress in biosensing systems: A review. *Trends Environ. Anal. Chem.* **2020**, *26*, e00087. [CrossRef]
3. Presutti, D.; Agarwal, T.; Zarepour, A.; Celikkin, N.; Hooshmand, S.; Nayak, C.; Ghomi, M.; Zarrabi, A.; Costantini, M.; Behera, B.; et al. Transition Metal Dichalcogenides (TMDC)-Based Nanozymes for Biosensing and Therapeutic Applications. *Materials* **2022**, *15*, 337. [CrossRef] [PubMed]
4. Wu, J.; Wang, X.; Wang, Q.; Lou, Z.; Li, S.; Zhu, Y.; Qin, L.; Wei, H. Nanomaterials with enzyme-like characteristics (nanozymes): Next-generation artificial enzymes (II). *Chem. Soc. Rev.* **2019**, *48*, 1004–1076. [CrossRef] [PubMed]
5. Huang, Y.; Ren, J.; Qu, X. Nanozymes: Classification, Catalytic Mechanisms, Activity Regulation, and Applications. *Chem. Rev.* **2019**, *119*, 4357–4412. [CrossRef] [PubMed]
6. Liu, Q.; Zhang, A.; Wang, R.; Zhang, Q.; Cui, D. A Review on Metal- and Metal Oxide-Based Nanozymes: Properties, Mechanisms, and Applications. *Nano Micro Lett.* **2021**, *13*, 154. [CrossRef] [PubMed]
7. Stasyuk, N.; Smutok, O.; Demkiv, O.; Prokopiv, T.; Gayda, G.; Nisnevitch, M.; Gonchar, M. Synthesis, Catalytic Properties and Application in Biosensorics of Nanozymes and Electronanocatalysts: A Review. *Sensors* **2020**, *20*, 4509. [CrossRef] [PubMed]
8. Neumann, B.; Wollenberger, U. Electrochemical Biosensors Employing Natural and Artificial Heme Peroxidases on Semiconductors. *Sensors* **2020**, *20*, 3692. [CrossRef] [PubMed]
9. Demkiv, O.; Stasyuk, N.; Serkiz, R.; Gayda, G.; Nisnevitch, M.; Gonchar, M. Peroxidase-Like Metal-Based Nanozymes: Synthesis, Catalytic Properties, and Analytical Application. *Appl. Sci.* **2021**, *11*, 777. [CrossRef]
10. Gayda, G.Z.; Demkiv, O.M.; Gurianov, Y.; Serkiz, R.Y.; Klepach, H.M.; Gonchar, M.V.; Nisnevitch, M. "Green" Prussian Blue Analogues as Peroxidase Mimetics for Amperometric Sensing and Biosensing. *Biosensors* **2021**, *11*, 193. [CrossRef] [PubMed]
11. Komkova, M.A.; Pasquarelli, A.; Andreev, E.A.; Galushin, A.A.; Karyakin, A.A. Prussian Blue modified boron-doped diamond interfaces for advanced H_2O_2 electrochemical sensors. *Electrochim. Acta* **2020**, *339*, 135924. [CrossRef]
12. Keihan, A.H.; Karimi, R.R.; Sajjadi, S. Wide dynamic range and ultrasensitive detection of hydrogen peroxide based on beneficial role of gold nanoparticles on the electrochemical properties of Prussian blue. *J. Electroanal. Chem.* **2020**, *862*, 114001. [CrossRef]
13. Tripathi, A.; Harris, K.D.; Elias, A.L. High surface area nitrogen-functionalized Ni nanozymes for efficient peroxidase-like catalytic activity. *PLoS ONE* **2021**, *16*, e0257777. [CrossRef] [PubMed]
14. Fu, Z.; Zeng, W.; Cai, S.; Li, H.; Ding, J.; Wang, C.; Chen, Y.; Han, N.; Yang, R. Porous Au@Pt nanoparticles with superior peroxidase-like activity for colorimetric detection of spike protein of SARS-CoV-2. *J. Colloid Interface Sci.* **2021**, *604*, 113–121. [CrossRef]
15. Stasyuk, N.; Gayda, G.; Demkiv, O.; Darmohray, L.; Gonchar, M.; Nisnevitch, M. Amperometric biosensors for L-arginine determination based on L-arginine oxidase and peroxidase-like nanozymes. *Appl. Sci.* **2021**, *11*, 7024. [CrossRef]
16. Stine, K.J.; Jefferson, K.; Shulga, O.V. Nanoporous gold for enzyme immobilization. *Methods Mol. Biol.* **2017**, *1504*, 37–60. [PubMed]
17. Bhattarai, J.K.; Neupane, D.; Nepal, B.; Mikhaylov, V.; Demchenko, A.V.; Stine, K.J. Preparation, Modification, Characterization, and Biosensing Application of Nanoporous Gold Using Electrochemical Techniques. *Nanomaterials* **2018**, *8*, 171. [CrossRef] [PubMed]
18. Klepach, H.M.; Zakalskiy, A.E.; Zakalska, O.M.; Gayda, G.Z.; Smutok, O.V.; Gonchar, M.V. Alcohol Oxidase from the Methylotrophic Yeast *Ogataea polymorpha*: Isolation, Purification, and Bioanalytical Application. *Flavins Flavoproteins* **2021**, *2280*, 231–248.
19. Stasyuk, N.E.; Smutok, O.V.; Zakalskiy, A.E.; Zakalska, O.M.; Gonchar, M.V. Methylamine-Sensitive Amperometric Biosensor Based on (His)$_6$-Tagged *Hansenula polymorpha* Methylamine Oxidase Immobilized on the Gold Nanoparticles. *Biomed. Res. Int.* **2014**, *2014*, 480498. [CrossRef] [PubMed]
20. Demkiv, O.; Stasyuk, N.; Gayda, G.; Gonchar, M. Highly sensitive amperometric sensor based on laccase-mimicking metal-based hybrid nanozymes for adrenaline analysis in pharmaceuticals. *Catalysts* **2021**, *11*, 1510. [CrossRef]

Abstract

Use of Fluorescent Yeast-Based Biosensors for Evaluation of the Binding Affinities of New Steroid Hormone and Bile Acid Derivatives for Select Steroid Receptors [†]

Sofija S. Bekić [1,*], Ivana Kuzminac [1], Srđan Bjedov [1], Jovana Ajduković [1], Marina Savić [1], Edward Petri [2] and Anđelka Ćelić [2]

[1] Department of Chemistry, Biochemistry and Environmental Protection, Faculty of Sciences, University of Novi Sad, 21000 Novi Sad, Serbia; ivana.kuzminac@dh.uns.ac.rs (I.K.); srdjan.bjedov@dh.uns.ac.rs (S.B.); jovana.ajdukovic@dh.uns.ac.rs (J.A.); marina.savic@dh.uns.ac.rs (M.S.)

[2] Department of Biology and Ecology, Faculty of Sciences, University of Novi Sad, 21000 Novi Sad, Serbia; edward.petri@dbe.uns.ac.rs (E.P.); andjelka.celic@dbe.uns.ac.rs (A.Ć.)

* Correspondence: sofija.bekic@dh.uns.ac.rs

† Presented at the 2nd International Electronic Conference on Biosensors, 14–18 February 2022; Available Online: https://sciforum.net/event/IECB2022.

Citation: Bekić, S.S.; Kuzminac, I.; Bjedov, S.; Ajduković, J.; Savić, M.; Petri, E.; Ćelić, A. Use of Fluorescent Yeast-Based Biosensors for Evaluation of the Binding Affinities of New Steroid Hormone and Bile Acid Derivatives for Select Steroid Receptors. *Eng. Proc.* **2022**, *16*, 4. https://doi.org/10.3390/IECB2022-12282

Academic Editors: Giovanna Marrazza and Sara Tombelli

Published: 15 February 2022

Publisher's Note: MDPI stays neutral with regard to jurisdictional claims in published maps and institutional affiliations.

Copyright: © 2022 by the authors. Licensee MDPI, Basel, Switzerland. This article is an open access article distributed under the terms and conditions of the Creative Commons Attribution (CC BY) license (https://creativecommons.org/licenses/by/4.0/).

Abstract: Biosensors developed in yeast cells represent an attractive research area in biomedicine because they allow for the detection of molecules of various structures and biological activities, economically and simply, without the use of harmful radioactive materials. We focused our attention on the identification of androgen, glucocorticoid and estrogen receptor α/β ligands using fluorescent biosensors in yeast. Identification of compounds that modulate the activity of androgen (AR) or estrogen receptors (ER) is one of the major goals in the design of new treatments of hormone-dependent cancers. Similarly, glucocorticoid receptor (GR) ligands are used to treat autoimmune and inflammatory diseases, but due to a large number of side effects and drug resistance, great effort has been directed to finding new modulators. In this study, ligand-binding domains (LBDs) of AR, ERα, ERβ or GR fused with yellow fluorescent protein (YFP) were expressed in *Saccharomyces cerevisiae*. Recombinant yeast cells were treated with tested steroid hormone or bile acid derivatives, and, due to the fluorescence resonance energy transfer phenomenon following ligand binding, relative binding affinities were quantified fluorometrically. Our results show that some of the tested compounds have moderate to high binding affinity for particular steroid receptors, similar to natural ligands, while the affinities of other compounds were low or negligible. To elucidate the mechanisms of action for these compounds, additional experiments are necessary, and to better understand the molecular interactions within the ligand-binding pocket of the receptor, molecular docking analysis can be conducted. In summary, the yeast-based biosensors used in this work have proven to be very useful for in vitro screening of novel anticancer and anti-inflammatory drug candidates, as well as for the elimination of compounds that do not deserve further attention and resources due to their lack of desired bioactivities.

Keywords: steroid receptor; cancer; ligand; modified steroid; bile acid; biosensor; FRET

Supplementary Materials: The following supporting information can be downloaded at: https://www.mdpi.com/article/10.3390/IECB2022-12282/s1.

Author Contributions: Conceptualization, A.Ć. and E.P.; methodology, S.S.B.; software, S.S.B.; validation, A.Ć., E.P. and S.S.B.; formal analysis, S.S.B.; investigation, S.S.B., I.K., S.B., J.A. and M.S.; resources, A.Ć., E.P. and M.S.; data curation, S.S.B. and A.Ć.; writing—original draft preparation, S.S.B.; writing—review and editing, A.Ć. and E.P.; visualization, S.S.B.; supervision, A.Ć. and E.P.; project administration, M.S.; funding acquisition, M.S. All authors have read and agreed to the published version of the manuscript.

Funding: This research was funded by the Provincial Secretariat for Higher Education and Scientific Research of the Autonomous Province of Vojvodina [Project: New steroid derivatives—potential chemotherapeutics, No. 142-451-2667/2021].

Institutional Review Board Statement: Not applicable.

Informed Consent Statement: Not applicable.

Conflicts of Interest: The authors declare no conflict of interest.

Proceeding Paper

Simultaneous Detection of *Salmonella typhimurium* and *Escherichia coli* O157:H7 in Drinking Water with Mach–Zehnder Interferometers Monolithically Integrated on Silicon Chips [†]

Michailia Angelopoulou [1,*], **Panagiota Petrou** [1], **Konstantinos Misiakos** [2], **Ioannis Raptis** [2] **and Sotirios Kakabakos** [1,*]

1   Immunoassays–Immunosensors Lab, Institute of Nuclear & Radiological Sciences & Technology, Energy & Safety, NCSR "Demokritos", 15341 Aghia Paraskevi, Greece; ypetrou@rrp.demokritos.gr

2   Institute of Nanoscience & Nanotechnology, NCSR "Demokritos", 15341 Aghia Paraskevi, Greece; k.misiakos@inn.demokritos.gr (K.M.); i.raptis@inn.demokritos.gr (I.R.)

*   Correspondence: mikangel@ipta.demokritos.gr (M.A.); skakab@rrp.demokritos.gr (S.K.); Tel.: +30-2106503819 (M.A. & S.K.)

†   Presented at the 2nd International Electronic Conference on Biosensors, 14–18 February 2022; Available Online: https://sciforum.net/event/IECB2022.

Abstract: The consumption of water contaminated with bacteria can lead to foodborne disease outbreaks. For this reason, the development of rapid and sensitive analytical methods for bacteria detection is of primary importance for public health protection. Here, a miniaturized immunosensor based on Mach–Zehnder Interferometry for the simultaneous, real-time determination of *S. typhimurium* and *E. coli* in drinking water is presented. For the assay, mixtures of bacteria solutions with anti-bacteria-specific antibodies were run over the chip, followed by biotinylated anti-species-specific antibody and streptavidin solutions. The assay was fast (10 min), accurate, sensitive (LOD: 3×10^2 cfu/mL for S. *typhimurium*; 2×10^2 cfu/mL for *E. coli*) and reproducible. The analytical characteristics achieved combined with the small chip size make the proposed biosensor suitable for on-site bacteria determination in drinking water samples.

Keywords: bacteria; *Salmonella typhimurium*; *Escherichia coli*; Mach–Zehnder interferometry; immunosensor

Citation: Angelopoulou, M.; Petrou, P.; Misiakos, K.; Raptis, I.; Kakabakos, S. Simultaneous Detection of *Salmonella typhimurium* and *Escherichia coli* O157:H7 in Drinking Water with Mach–Zehnder Interferometers Monolithically Integrated on Silicon Chips. *Eng. Proc.* **2022**, *16*, 5. https://doi.org/10.3390/IECB2022-12269

Academic Editors: Giovanna Marrazza and Sara Tombelli

Published: 14 February 2022

Publisher's Note: MDPI stays neutral with regard to jurisdictional claims in published maps and institutional affiliations.

Copyright: © 2022 by the authors. Licensee MDPI, Basel, Switzerland. This article is an open access article distributed under the terms and conditions of the Creative Commons Attribution (CC BY) license (https://creativecommons.org/licenses/by/4.0/).

1. Introduction

The consumption of food and water contaminated with pathogens is of global interest as it leads to 48×10^6 infections annually, resulting in 128,000 hospitalizations and 3000 deaths [1]. According to CDC, the estimated incidents of foodborne illness caused by 31 pathogenic bacteria in the US amount to a total of 9 million cases per year, from which 20% is attributed to *Salmonella* spp., *Escherichia coli* O157:H7, *Staphylococcus aureus*, Clostridium perfringens, *Campylobacter* spp. and *Shigella* spp. [2]. Among them, *Salmonella typhimurium* (*S. typhimurium*) and *Escherichia coli* O157:H7 (*E. coli* O157:H7) are both facultatively anaerobic, rod-shaped, Gram-negative bacteria, belonging to the Enterobacteriaceae family and are most frequently associated with foodborne illnesses. The ingestion of *S. typhimurium* causes fever, nausea, diarrhea, stomach discomfort, vomiting, dehydration and weakness, while *E. coli* may cause, on top of the aforementioned symptoms, potentially life-threatening complications known as hemolytic uremic syndrome and hemorrhagic colitis. In both cases, the clinical symptoms may last from 5 to 7 days [3,4]. The number of outbreaks due to *S. typhimurium* and *E. coli* O157:H7 infections in combination with the economic loss associated with these foodborne illnesses due to medical costs, loss of work hours and product recalls have imposed the need for rapid diagnostic methods for pathogen detection [5].

The conventional methods for bacteria detection and identification are based on culturing and plating. Those methods are reliable but include several steps, such as pre-enrichment, selective enrichment, isolation and confirmation through biochemical and serological tests, which are rather time consuming and require at least 5–7 days to complete. In order to shorten the analysis time to 2–4 days, ELISA- and DNA-based methods have been employed for bacteria identification, thus replacing the selective plating steps [6,7].

In recent years, biosensors based on electrochemical, piezoelectric or optical transducers are gaining ground in foodborne bacteria detection. Concerning electrochemical immunosensors, devices employing amperometric, potentiometric, impedemetric and conductimetric detection principles have been developed for detection of bacteria [8–10]. Although these sensors claim inexpensive analysis and potential for miniaturization, they often require labels for signal enhancement to improve detection limits. Similarly, immunosensors based on piezoelectric phenomena are capable of label-free detection but they lack in sensitivity [11]. On the other hand, optical biosensors utilizing different transduction principles, such as light absorbance, SPR, fluorescence, light polarization and Raman scattering, are powerful tools for foodborne bacteria detection [12]. Optical detection provides several advantages over other transduction principles, such as less interference from the sample and ability for direct determination of pathogens in complex matrices with minimal sample treatment. Although SPR biosensors are widely used for label-free bacteria detection, their limit of detection is usually higher than 10^3 cfu/mL, [13,14]. Among the label-free biosensors, interferometric ones are the most promising bacteria detection systems, as they offer high sensitivity and multiplexing capability for real-time determinations. Recently, a bi-modal interferometric sensor, an interferometric reflectance imaging sensor and a microcavity in-line Mach–Zehnder interferometer have been employed for the detection of *E. coli* with detection limits of 40, 2.2 and 100 cfu/mL, respectively [15–17]. Moreover, another interferometric sensor based on white light reflectance spectroscopy has been developed for the detection of *S. typhimurium* in drinking water samples, exhibiting a detection limit of 320 cfu/mL [18].

In this work, a label-free optical immunosensor based on arrays of Mach–Zehnder interferometers (MZIs) monolithically integrated onto silicon chips, which are appropriate for multi-analyte determinations, is employed. The MZIs chips have been successfully utilized for the determination of allergens and mycotoxins in foodstuffs, as well as for the detection of goat milk and PDO cheeses adulteration with bovine milk [19–22]. Here, the MZIs chips are employed, for the first time, for the simultaneous determination of *S. typhimurium* and *E. coli* in drinking water samples. The detection of bacteria was based on the competitive immunoassay principle through biofunctionalization of the sensing arm of the MZIs with the lipopolysaccharides (LPS) of the two bacteria (Figure 1). Biomolecular reactions on the LPS modified sensing arm change the effective refractive index, causing a blue shift of the interference spectrum. The spectral shift is transformed to phase shift, and the signal is expressed in radians. Several assay parameters were optimized aiming for fast and sensitive simultaneous determination of both bacteria in drinking water.

Figure 1. Three-dimensional schematic of assay configuration for bacteria detection using the MZI sensor.

2. Experimental Section

2.1. Materials

Salmonella enterica serovar *typhimurium* (*S. typhimurium*, ATCC 14028) and *Escherichia coli* O157:H7 (*E. coli* O157:H7, NCTC 12900) were kindly provided from Delta Foods S.A. (Athens, Greece). *E. coli* LPS was obtained from Creative Diagnostics (Upton, NY, USA). The goat polyclonal antibody against *E. coli* LPS was from Kirkegaard & Perry Lab Inc. (Gaithersburg, MD, USA). The rabbit polyclonal antibody against *S. typhimiurim* LPS, donkey anti-goat IgG antibody and donkey anti-rabbit IgG antibody were purchased from Bio-Rad (Watford, UK). Salmonella LPS, bovine serum albumin (BSA) and 3-aminopropyltriethoxysilane (APTES) were purchased from Sigma-Aldrich (Darmstadt, Germany). Streptavidin was from Thermo-Scientific (Waltham, MA, USA). The water used in the study was double distilled. Donkey anti-goat IgG and donkey anti-rabbit IgG antibodies were biotinylated according to a previously published protocol [22].

2.2. Chip Fabrication and Signal Processing

Fabrication of the chips was performed following mainstream silicon technology as described previously [21,22]. The chip consists of an array of ten silicon nitride MZIs, each one of them coupled with a respective silicon LED. The ten LEDs are serially turned on and off using a multiplexer. The chip is covered by a silicon oxide cladding layer that was selectively removed from a 600 μm long area over the sensing arm of each MZI to allow for interaction of the waveguided photons with the spotted biomolecules onto the sensing arm. The ten MZIs converge in a single output at the edge of the chip where the transmitted light is collected by an external spectrometer (QE65000, Ocean Optics, Orlando, FL, USA). The spectral shifts caused by the immunoreactions over the sensing arm of the MZIs are continuously recorded and converted to phase shifts through discrete Fourier transform.

2.3. Chemical and Biological Functionalization of the Chip

For the chemical activation, the chips were cleaned and hydrophilized through O_2 plasma treatment for 30 s. Then, they were immersed for 2 min in a 0.5% (*v/v*) APTES solution, rinsed, dried under nitrogen stream and heated at 120 °C for 20 min. The biological activation of the chips was performed using the BioOdyssey Calligrapher Mini Arrayer. Hence, 3 MZIs per chip were spotted with 100 μg/mL of *S. typhimurium* LPS solution, 4 MZIs with 50 μg/mL of *E. coli* LPS solution and the remaining 3 with 100 μg/mL of BSA solution for the determination of non-specific binding. After the completion of spotting, the chips were incubated overnight at 4 °C in humidity chamber. Then, the biofunctionalized chips were washed and incubated for 1 h in 1% (*w/v*) BSA in 0.1 M $NaHCO_3$ solution to block the non-specific binding sites on the sensing arm, rinsed with water and dried under nitrogen stream.

2.4. Immunoassay for Bacteria Detection with MZI Immunosensor

The delivery of the samples over the chip surface was achieved through attachment of an appropriate microfluidic module onto the chip. Then, the chip was placed on a handling frame and inserted in the docking station of the measuring device. Prior to assay, calibrators/samples were mixed with the antibodies against *S. typhimurium* and *E. coli* LPS at 1:1 volume ratio and incubated for 30 min. After chip equilibration with assay buffer, 100 μL of these mixtures was run over the chip at a rate of 35 μL/min, followed by 100 μL of biotinylated anti-species-specific antibodies and 100 μL of streptavidin solution (Figure 1). After assay completion, a regeneration step was followed in order to remove the bound antibodies from the biofunctionalized surface and reuse the chip for the next sample. Thus, 100 μL of 0.05 M HCl solution and 100 μL of 0.05 M NaOH solution were pumped over the chip sequentially, and finally, 100 μL of assay buffer was flown for chip equilibration.

3. Results

3.1. Optimization of Assay Parameters

The detection of bacteria in drinking water samples was based on the competitive immunoassay principle. Thus, several parameters were optimized with respect to the maximum signal and the percent signal drop obtained for certain bacteria calibrators. Firstly, the optimum concentration of bacteria LPS for immobilization onto the sensor's surface was determined by running for 1 h zero calibrators over chips spotted with LPS solutions with concentrations ranging from 5 to 200 µg/mL. As shown in Figure 2a, indicatively for *E. coli* LPS, the signal increased and reached a maximum at a concentration of 50 µg/mL, whereas for concentrations higher than 100 µg/mL the signal started to decline. Furthermore, the sensor-to-sensor and the chip-to-chip signal variation was significantly improved (CV < 5%) when the concentration of LPS used for coating was equal or higher than 50 µg/mL compared to the signal variations obtained from chips spotted with lower concentrations. Thus, 50 µg/mL concentration of *E. coli* LPS was selected for further experimentation. In a similar way, the optimum concentration for immobilization regarding the *S. typhimurium* LPS was determined to be 100 µg/mL.

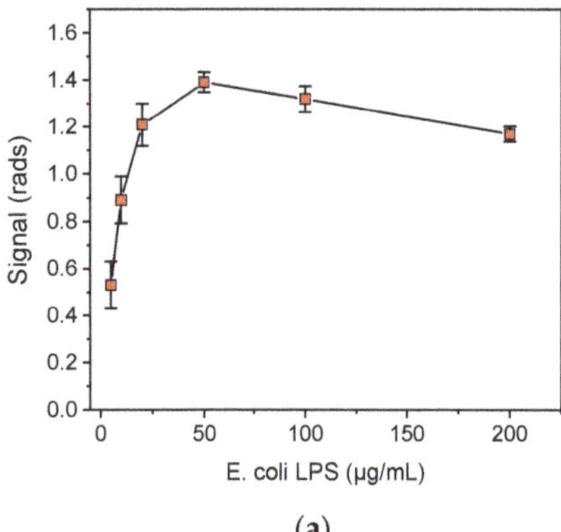

(a)

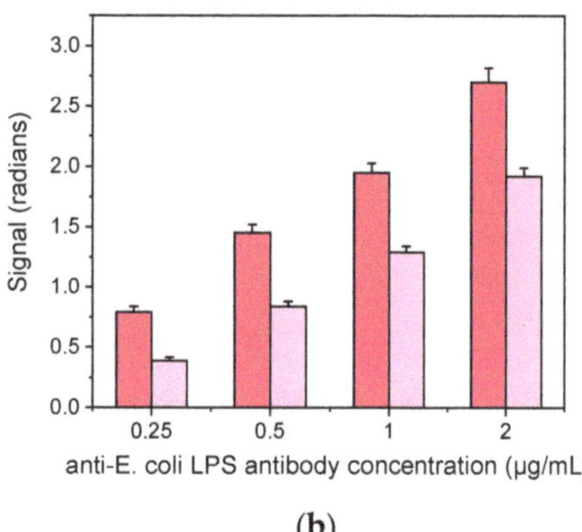

(b)

Figure 2. (a) Signal values for zero calibrator vs. *E. coli* LPS concentrations used for coating and (b) signals corresponding to zero calibrator (magenta columns) or an *E. coli* LPS calibrator with concentration of 0.025 µg/mL (purple columns) obtained using different concentrations of anti-*E. coli* antibody. Each point is the mean value of ten waveguides per chip ± SD.

Another parameter optimized was the concentration of the anti-bacteria antibodies in order to select the one providing adequate signal in combination with detection sensitivity. The concentrations tested ranged from 0.25 to 2 µg/mL for *E. coli* LPS, and 0.5 to 3 µg/mL for *S. typhimurium*. As presented in Figure 2b, indicatively for E. coli, adequate signal (≥1 rad) was achieved for anti-*E. coli* LPS concentrations ≥0.5 µg/mL. However, although higher signals were obtained by increasing the antibody concentration, the higher detection sensitivity was achieved for antibody concentration of 0.5 µg/mL. For *S. typhimurium*, the optimum antibody concentration was 1 µg/mL.

3.2. Matrix Effect

For the determination of bacteria in drinking water samples, the effect of tap water on the signal was investigated. For this reason, *E. coli* and *S. typhimurium* zero calibrators were prepared in assay buffer as well as in tap water. As shown in Figure 3, indicatively for *S. typhimurium*, the signal obtained from zero calibrator prepared in tap water was similar to that of the zero calibrator prepared in assay buffer. Moreover, the calibration curves obtained with calibrators prepared in both matrices were almost identical. Thus, the calibrators were prepared in assay buffer.

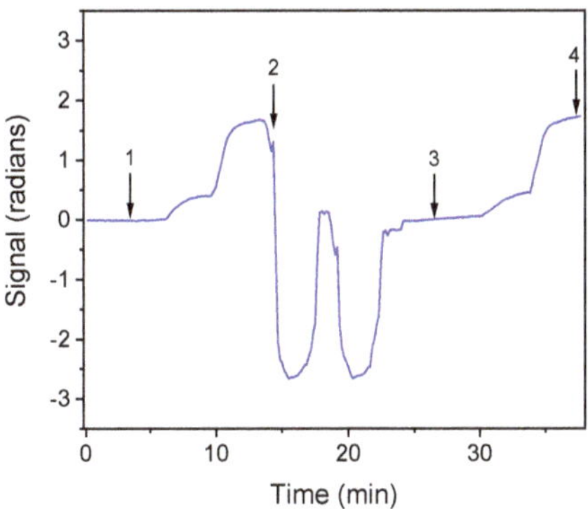

Figure 3. Real-time response obtained for *S. typhimurium* zero calibrator prepared in: assay buffer (arrow 1 to 2); regeneration and equilibration (arrow 2 to 3); and zero calibrator in tap water (arrow 3 to 4).

3.3. Analytical Characteristics and Calibration Curves Using the MZI Chip

In Figure 4, the calibration curves for *S. typhimurium* and *E. coli* LPS are provided. The dynamic range of *S. typhimurium* and *E. coli* LPS assays ranged from 0.005 to 1 μg/mL and from 0.005 to 0.5 μg/mL, respectively. The limit of detection of the assays was determined as the concentration corresponding to signal equal to -3SD of the mean zero calibrator signals (28 replicate values from 4 chips; 7 MZIs per chip) and was 0.004 μg/mL for both bacteria. Moreover, the limit of detection was 3×10^2 cfu/mL and 2×10^2 cfu/mL for *S. typhimurium* and *E. coli*, respectively. The accuracy of the assay was also determined through recovery experiments. For this reason, tap water was spiked with three different bacteria concentrations. The recovery values ranged from 91 to 112%, indicating the high accuracy of the assay performed using the MZI chip. The repeatability of the assay was determined using tap water samples spiked with four different concentrations of the bacteria. The intra-assay coefficients of variation (CVs) were calculated after repetitive measurements of the tap water samples during the same day, whereas the inter-assay CVs were determined by measuring the tap water samples in seven different days in a period of one month and were less than 5% and 7%, respectively.

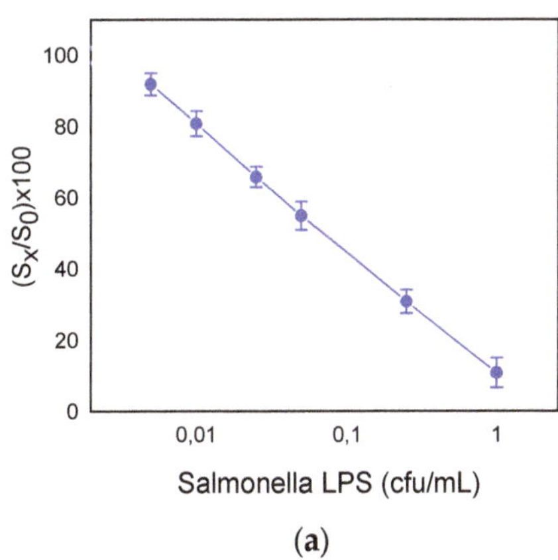

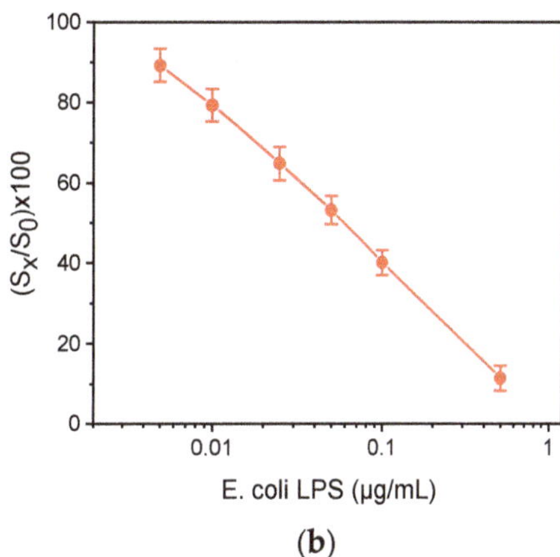

(a)

(b)

Figure 4. Calibration curves of (**a**) *S. typhimurium* LPS and (**b**) *E. coli* LPS. $(S_x/S_0) \times 100$ represents the percent ratio of each calibrator signal (S_x) to the zero calibrator signal (S_0). Each point is the mean value of seven waveguides per chip $\pm$ SD.

4. Conclusions

The simultaneous determination of *S. typhimurium* and *E. coli* in drinking water using the MZI immunosensor chips was presented. The sensor provided real-time detection of the two bacteria in 10 min, employing a three-step assay configuration. The assay was accurate, repeatable and sensitive with detection limits at the order of 10^2 cfu/mL. Thus, it is expected that the proposed sensor could find wide application in Drinking Water Distribution System and in low-resources environment for the fast on-site monitoring of bacteria.

Author Contributions: Conceptualization, M.A., P.P., I.R., K.M. and S.K.; M.A. and P.P.; formal analysis, M.A.; investigation, M.A.; resources, K.M. and I.R.; data curation, M.A. and S.K.; writing—original draft preparation, M.A.; writing—review and editing, P.P., K.M. and I.R.; visualization, M.A.; supervision, P.P. and S.K.; project administration, S.K.; funding acquisition, M.A. and S.K. All authors have read and agreed to the published version of the manuscript.

Funding: M.A. was supported by the program of Industrial Scholarships of Stavros Niarchos Foundation.

Institutional Review Board Statement: Not applicable.

Informed Consent Statement: Not applicable.

Data Availability Statement: The data presented in this study are available on request from the corresponding author.

Conflicts of Interest: The authors declare no conflict of interest.

References

1. Centers for Disease Control and Prevention. Available online: https://www.cdc.gov/foodborneburden/estimates-overview.html (accessed on 30 March 2022).
2. Scallan, E.; Hoekstra, R.M.; Angulo, F.J.; Tauxe, R.V.; Widdowson, M.A.; Roy, S.L.; Jones, J.L.; Griffin, P.M. Foodborne illness acquired in the United States—Major pathogens. *Emerg. Infect. Dis.* **2011**, *17*, 7–15. [CrossRef] [PubMed]
3. Inbaraj, B.S.; Chen, B. Nanomaterial-based sensors for detection of foodborne bacterial pathogens and toxins as well as pork adulteration in meat products. *J. Food Drug Anal.* **2016**, *24*, 15–28. [CrossRef] [PubMed]
4. Waswa, J.W.; Debroy, C.; Irudayaraj, J. Rapid detection of Salmonella enteritidis and Escherichia coli using Surface Plasmon Resonance biosensor. *J. Food Process Eng.* **2006**, *29*, 373–385. [CrossRef]
5. Xu, L.; Bai, X.; Bhunia, A.K. Current state of development of biosensors and their application in foodborne pathogen detection. *J. Food Prot.* **2021**, *84*, 1213–1227. [CrossRef] [PubMed]
6. Walker, D.I.; McQuillan, J.; Taiwo, M.; Parks, R.; Stenton, C.A.; Morgan, H.; Mowlem, M.C.; Lees, D.N. A highly specific Escherichia coli qPCR and its comparison with existing methods for environmental waters. *Water Res.* **2017**, *126*, 101–110. [CrossRef] [PubMed]
7. Wang, W.; Liu, L.; Song, S.; Tang, L.; Kuang, H.; Xu, C. A highly sensitive ELISA and immunochromatographic strip for the detection of *Salmonella typhimurium* in milk samples. *Sensors* **2015**, *15*, 5281–5292. [CrossRef] [PubMed]
8. Zelada-Guillén, G.A.; Bhosale, S.V.; Riu, J.; Rius, F.X. Real-time potentiometric detection of bacteria in complex samples. *Anal. Chem.* **2010**, *82*, 9254–9260. [CrossRef]
9. Xu, M.; Wang, R.; Li, Y. An electrochemical biosensor for rapid detection of *E. coli* O157:H7 with highly efficient bifunctional glucose oxidase-polydopamine nanocomposites and Prussian blue modified screen-printed interdigitated electrodes. *Analyst* **2016**, *141*, 5441–5449. [CrossRef]
10. Subjakova, V.; Oravczova, V.; Tatarko, M.; Hianik, T. Advances in electrochemical aptasensors and immunosensors for detection of bacterial pathogens in food. *Electrochim. Acta* **2021**, *389*, 138724. [CrossRef]
11. Salam, F.; Uludag, Y.; Tothill, I.E. Real-time and sensitive detection of *Salmonella typhimurium* using an automated quartz crystal microbalance (QCM) instrument with nanoparticles amplification. *Talanta* **2013**, *115*, 761–767. [CrossRef]
12. Terry, L.A.; White, S.F.; Tigwell, L.J. The application of biosensors to fresh produce and the wider food industry. *J. Agric. Food Chem.* **2005**, *53*, 1309–1316. [CrossRef] [PubMed]
13. Xu, Y.; Luo, Z.; Chen, J.; Huang, Z.; Wang, X.; An, H.; Duan, Y. Novel W-shaped fiber-optic probe-based localized surface Plasmon resonance biosensor for real-time detection of *Salmonella typhimurium*. *Anal. Chem.* **2018**, *90*, 13640–13646. [CrossRef] [PubMed]
14. Kaushika, S.; Tiwaria, U.K.; Pala, S.S.; Sinhaa, R.K. Rapid detection of Escherichia coli using fiber optic surface plasmon resonance immunosensor based on biofunctionalized Molybdenum disulfide (MoS2) nanosheets. *Biosens. Bioelectron.* **2019**, *126*, 501–509. [CrossRef]
15. Maldonado, J.; González-Guerrero, A.B.; Domínguez, C.; Lechuga, L.M. Label-free bimodal waveguide immunosensor for rapid diagnosis of bacterial infections in cirrhotic patients. *Biosens. Bioelectron.* **2016**, *85*, 310–316. [CrossRef] [PubMed]

16. Zaraee, N.; Kanik, F.E.; Bhuiya, A.M.; Gong, E.S.; Geib, M.T.; Ünlü, N.L.; Ozkumur, A.Y.; Dupuis, J.R.; Ünlü, M.S. Highly sensitive and label-free digital detection of whole cell E. coli with Interferometric Reflectance Imaging. *Biosens. Bioelectron.* **2020**, *162*, 112258. [CrossRef] [PubMed]

17. Janik, M.; Koba, M.; Celebanska, A.; Bock, W.J.; Smietana, M.; Live, E. coli bacteria label-free sensing using a microcavity in-line Mach-Zehnder interferometer. *Sci. Rep.* **2018**, *8*, 17176. [CrossRef]

18. Angelopoulou, M.; Tzialla, K.; Voulgari, A.; Dikeoulia, M.; Raptis, I.; Kakabakos, S.E.; Petrou, P. Rapid Detection of *Salmonella typhimurium* in drinking water by a White Light Reflectance Spectroscopy immunosensor. *Sensors* **2021**, *21*, 2683. [CrossRef]

19. Angelopoulou, M.; Petrou, P.S.; Makarona, E.; Haasnoot, W.; Moser, I.; Jobst, G.; Goustouridis, D.; Lees, M.; Kalatzi, K.; Raptis, I.; et al. Ultrafast multiplexed-allergen detection through advanced fluidic design and monolithic interferometric silicon chips. *Anal. Chem.* **2018**, *90*, 9559–9567. [CrossRef]

20. Pagkali, V.; Petrou, P.S.; Makarona, E.; Peters, J.; Haasnoot, W.; Jobst, G.; Moser, I.; Gajos, K.; Budkowski, A.; Economou, A.; et al. Simultaneous determination of aflatoxin B 1, fumonisin B 1 and deoxynivalenol in beer samples with a label-free monolithically integrated optoelectronic biosensor. *J. Hazard. Mater.* **2018**, *359*, 445–453. [CrossRef]

21. Angelopoulou, M.; Botsialas, A.; Salapatas, A.; Petrou, P.S.; Haasnoot, W.; Makarona, E.; Gerhard, J.; Goustouridis, D.; Siafaka-Kapadai, A.; Raptis, I.; et al. Assessment of goat milk adulteration with a label-free monolithically integrated optoelectronic biosensor. *Anal. Bioanal. Chem.* **2015**, *407*, 3995–4004. [CrossRef]

22. Angelopoulou, M.; Petrou, P.S.; Raptis, I.; Misiakos, K.; Livaniou, E.; Makarona, E.; Kakabakos, S. Rapid detection of mozzarella and feta cheese adulteration with cow milk through a silicon photonic immunosensor. *Analyst* **2021**, *146*, 529. [CrossRef] [PubMed]

engineering proceedings

Proceeding Paper

Application of Optical and Acoustic Methods for the Detection of Bacterial Pathogens Using DNA Aptamers as Receptors †

Ivan Piovarci [1,*], Judit Süle [2], Michailia Angelopoulou [3], Panagiota Petrou [3], Leda Bousiakou [4], Sotirios Elias Kakabakos [3] and Tibor Hianik [1]

[1] Faculty of Mathematics, Physics and Informatics, Comenius University, Mlynska dolina F1, 842 48 Bratislava, Slovakia; tibor.hianik@fmph.uniba.sk
[2] Hungarian Dairy Research Institute Ltd., Lucsony utca 24, 9200 Mosonmagyaróvár, Hungary; jsule@mtki.hu
[3] National Centre for Scientific Research "Demokritos", Patr. Gregoriou E & 27 Neapoleos Str., 15341 Agia Paraskevi, Greece; mikangel@ipta.demokritos.gr (M.A.); ypetrou@rrp.demokritos.gr (P.P.); skakab@rrp.demokritos.gr (S.E.K.)
[4] IMD Laboratories Co., R&D Section, Lefkippos Technology Park, NCSR Demokritos, Agia Paraskeyi, P.O. Box 60037, 15130 Athens, Greece; leda07@hotmail.com
* Correspondence: piovarci.i@gmail.com
† Presented at the 2nd International Electronic Conference on Biosensors, 14–18 Feb 2022; Available Online: https://sciforum.net/event/IECB2022.

check for updates

Citation: Piovarci, I.; Süle, J.; Angelopoulou, M.; Petrou, P.; Bousiakou, L.; Kakabakos, S.E.; Hianik, T. Application of Optical and Acoustic Methods for the Detection of Bacterial Pathogens Using DNA Aptamers as Receptors. *Eng. Proc.* **2022**, *16*, 6. https://doi.org/10.3390/IECB2022-12268

Academic Editors: Giovanna Marrazza and Sara Tombelli

Published: 14 February 2022

Publisher's Note: MDPI stays neutral with regard to jurisdictional claims in published maps and institutional affiliations.

Copyright: © 2022 by the authors. Licensee MDPI, Basel, Switzerland. This article is an open access article distributed under the terms and conditions of the Creative Commons Attribution (CC BY) license (https://creativecommons.org/licenses/by/4.0/).

Abstract: Bacterial contamination in food is real and presents a valid danger to human health. Therefore, we focused on the detection of *Escherichia coli* and *Listeria innocua* with optical and acoustic methods. In both methods, we used specific DNA aptamers as receptors. For the optical method, we modified gold nanoparticles (AuNPs) with aptamers and analyzed the interaction of AuNPs with bacteria by measurement of the changes in the absorbance spectrum. We also applied white light reflectometry to measure changes in thickness on a silicon chip modified with aminylated aptamer through silica chemistry. We also used quartz crystal microbalance (QCM) in multiharmonic mode. In this case, the thiolated aptamers were chemisorbed at the gold layer of the quartz crystal and the changes in resonant frequency were measured following the addition of bacteria. The limit of detection (LOD) of the optical method using AuNPs was estimated to be 10^5 CFU/mL of *Listeria monocytogenes*. For the reflectometric method, we were able to detect *E. Coli* at concentrations around 2×10^4 CFU/mL. Using TSM, we analyzed the viscoelastic properties of the aptamer layers when they formed at the surface.

Keywords: *Escherichia coli*; *Listeria monocytogenes*; QCM; colorimetry; reflectometry

1. Introduction

Bacterial contamination in food presents a serious danger to human health. Every year, tons of food need to be thrown out and it is estimated that billions of people get sick from food poisoning, leading to deaths in some hundreds of thousand cases (mostly in children) [1]. Most of the bacterial contaminations can be traced to about 20–30 pathogenic bacteria [2,3]. In our work, we focused on the detection of *Escherichia coli* and *Listeria monocytogenes* with optical and acoustic methods. In both methods, we used specific DNA aptamers as receptors. Aptamers are single-stranded DNA or RNA that, in the correct environment, folds into structures, specifically binding to a target with constant of dissociation, K_d, around 10 nM or lower [4,5]. For the optical method, it is possible to modify gold nanoparticles (AuNPs) with aptamers and to study the interaction of AuNPs with bacteria by colorimetry. Aptamers also electrostatically interact with the AuNPs without specific binding, and it is possible to increase the stability of AuNPs with high ionic strength. The incubation of AuNPs with bacteria then removes the protecting aptamer and, after the addition of salts, causes AuNPs aggregation. This process can be measured by changes in absorbance spectra as a function of bacterial concentration. Another optical method used was white light reflectometry. This method measures changes in the thickness of the sensing layer deposited on a silicon chip modified with amino group-modified aptamer through silica chemistry. In this method, the white light from an optical wire illuminates the silicon surface and the reflected light is taken from the same wire to a spectrometer to measure the resulting spectrum and interference pattern of the reflected light. The mathematical model calculates changes in the apparent thickness of the sensing layer on a SiO_2 surface. We also used quartz crystal microbalance (QCM) in a multiharmonic mode for detection bacteria. In this case, the gold surface of the piezo crystal is modified by thiolated aptamers. Changes in resonant frequency and dissipation served as a signal are related to the interaction of bacteria with aptamers. It seems that the interaction of bacteria with the surface is not strictly due to the mass changes. Therefore, we also analyzed the changes in the viscoelastic properties of the sensing layer.

2. Materials and Methods

2.1. Chemicals and Materials

$HAuCl_4$, sodium citrate, phosphate buffer saline (PBS) tablets, TRIS-HCl, NaOH, BS3 linker, bovine serum albumin (BSA), (3-Aminopropyl)triethoxysilane (APTES), tris(2-carboxyethyl)phosphine (TCEP), NH_3, H_2O_2, H_2SO_4, acetone, and isopropyl alcohol were purchased from Sigma Aldrich (Darmstadt, Germany). DNA aptamers for *Listeria monocytogenes* and *E.coli* were purchased from Generi Biotech, Ltd. (Hradec Králové, Czech Republic).

2.2. Preparation of Gold Nanoparticles and Colorimetric Method

The gold nanoparticles (AuNPs) were prepared by the Turkevich method. Then, 100 mL of 0.25 mM $HAuCl_4$ was boiled by putting it on a hotplate set to 99.9 °C for around 20 min under continuous mixing with a magnetic stirrer. The top of the flask was covered with a Petri dish in order to protect the solution from external contaminants and to reduce a loss of solvent due to evaporation. After the solution was brought to boil, 3.4 mL of 39 mM trisodium citrate was added, so that the resultant molar ratio was 1:5 of $HAuCl_4$ to trisodium citrate. The solution first lost its yellow color and then darkened to violet color. This solution was maintained in boiling under mixing for around 15–20 min, after which the solution continually changed its color to red. We then removed the solution from the hot plate and left it to cool to room temperature. The AuNPs were stored in fridge at 4 °C. This method helped to obtain AuNPs with a diameter of approximately 5 nm.

The colorimetric experiments were performed by means of Thermo Scientific Genesys UV–vis spectrophotometer (Waltham, MA, USA). *Listeria monocytogenes* was prepared in Hungarian Dairy Research Institute (Mosonmagyarovar, Hungary) using standard microbiology methods. *Escherichia coli* O157:H7 was prepared in National Centre for Scientific Research "Demokritos" (Agia Paraskevi, Greece) in Petri dishes on agar using the standard microbiology method. Concentrations of both bacteria were determined by the plate counting method.

2.3. Preparation of Reflectometric Surface and Method of Detection

The surface of silicone SiO_2 chip was cleaned with acetone and isopropyl alcohol for 10 min during sonication. We then used acidic piranha for 20 min (1:1, H_2SO_4:H_2O_2) and then cleaned the surface with distilled water until dry under nitrogen. The surface was then incubated in 2% solution of APTES for 20 min. After that, it was dried with nitrogen and put into the oven at 120 °C for 20 min. Then, we added 80 μL of 10 μM aptamer with 8 μL of 600 μM BS3 linker for one hour. After this step, the surface was cleaned with distilled water and left to incubate in TRIS buffer overnight. For blocking of the naked surface, it was immersed into 1% BSA solution for one hour. The samples were measured in flow by a system of visible-near infrared light source (ThetaMetrisis S.A., Athens, Greece) with miniaturized USBE controlled spectrometer (Maya 2000 Pro; Ocean Insight, Orlando, FL,

USA) and a reflection probe (AVANTES Inc., Broomfield, CO, USA) consisting of seven optical fibers.

2.4. Preparation of QCM Surface and Method of Detection

The surface was cleaned using basic piranha (1:1:7, NH_3:H_2O_2:H_2O). We then applied sodium citrate buffer (250 mM, pH 3) in the flow. After stabilization, the aptamers in a concentration of 10 μM in a citrate buffer were added. We detected the samples in flow by using an injection pump. The QCM crystals (Total Frequency Control, Storrington, UK) were fitted in a quartz flow cell. The signal from the cell was taken with a SARK-110 (Seeed Studios, Shenzhen, China) antenna analyzer.

3. Results and Discussion

First, we explored colorimetry for detection of *Listeria monocytogenes*. We then determined the lowest NaCl concentration (80 mM) that causes the aggregation of AuNPs.

In the next series of the experiments, we incubated AuNPs for 15 min with different concentrations of DNA aptamers. The interaction of AuNPs with the aptamers protected them from aggregation in the presence of 200 mM NaCl. The spectrum of such AuNPs hardly changed with the use of aptamer concentration of 1 μM. In the next experiments, we incubated the 1 μM aptamer with different concentrations of *Listeria monocytogenes* for 40 min at 37 °C. The sample was then centrifuged for 10 min at 14,500 rpm. Bacteria bind to the aptamers in the sample. With increased concentration of bacteria, there is less aptamers in the supernatant after centrifugation. The supernatant was incubated with AuNPs for 15 min and the absorbance spectrum was measured immediately after adding 80 mM NaCl and 10 min after the addition of NaCl. Figure 1 shows the change in absorbance spectra after addition of NaCl for different concentrations of *L. monocytogenes*.

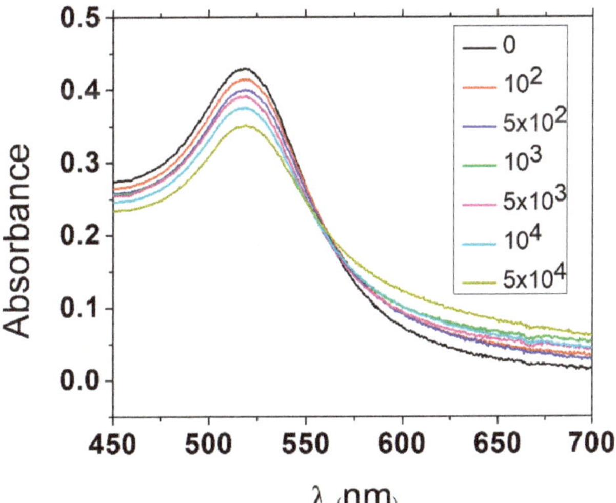

Figure 1. Change in absorbance spectra of gold nanoparticles 10 min. after addition of 80 mM NaCl for different concentrations of incubated *L. monocytogenes* in CFU/mL (See the legend).

Figure 2 shows the calibration curve consisting of change in absorbance (proportion of absorbance at 519 nm and 580 nm as peaks for non-aggregated and aggregated AuNPs, respectively) vs. concentration of bacteria. We calculated the limit of detection (LOD) of this assay as 5.5×10^3 CFU/mL. This method can quickly and easily detect bacteria with relatively high LOD. However, it is likely that the sensitivity of detection can be further improved by optimization of the conditions, such as salt concentration, as well as application of different salts, aptamer or AuNPs with another size.

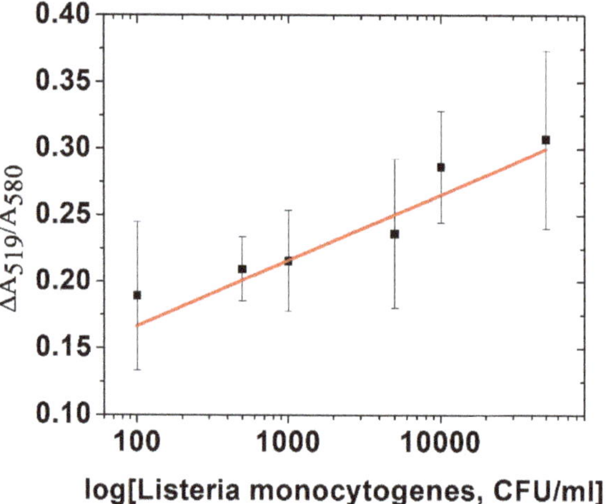

Figure 2. The plot of the ratio of absorbance at 519 nm and 580 nm (A519/A580) vs. concentration of *Listeria monocytogenes* in log scale. The results represent mean±SD obtained from 3 independent experiments. Red curve is linear fit performed by OriginPro version 7.5 (OriginLab Corporation, Northampton, MA, USA).

In the white light reflectometry method, we modified the silicon chip with an amino group-modified aptamers using APTES and a BS3 diester linker. Subsequently, we measured the reflection of white light from the surface of the silicon chip. After adding different concentrations of bacteria, we were able to monitor the change in the reflected spectrum and, using a mathematical model, we linked these changes to the variations in layer thickness. Typical kinetics related to changes in the thickness of naked silicon surface and those modified by aptamers (the naked SiO_2 surface was blocked by BSA) in the presence of *E. coli* are shown in Figure 3. An optical signal from bacteria can be seen for both surfaces; however, the kinetics for blocked surface were faster. After adding different concentrations of *E. coli* to the surface, we observed a 0.01–1 nm shift in the thickness depending on the concentration of bacteria starting from 2×10^2 CFU/mL (Figure 4).

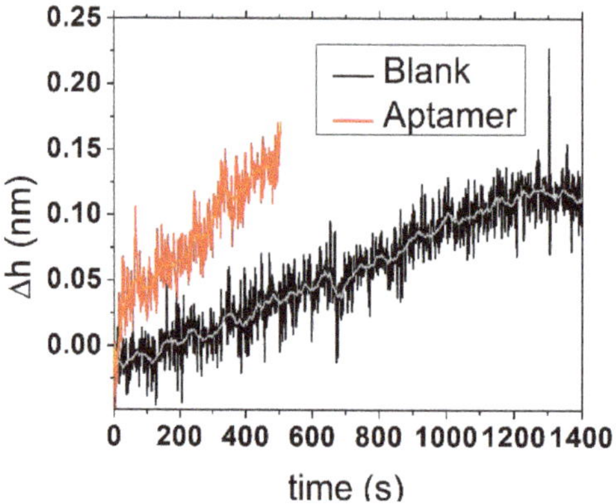

Figure 3. The kinetics of the thickness changes (Δh) obtained from reflectometry signal for SiO_2 surface with and without aptamers.

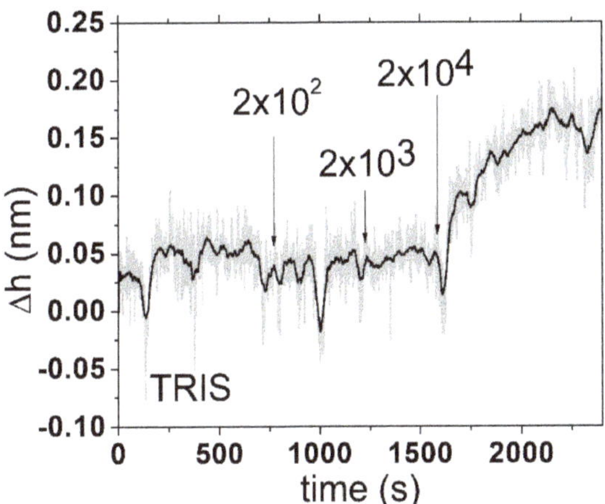

Figure 4. The kinetics of the thickness changes (Δh) obtained from reflectometric signal following the addition of *E. coli* (in CFU/mL) to a surface modified by DNA aptamers. The naked SiO_2 surface was blocked by BSA and washed by TRIS buffer. The addition of bacteria is shown by arrows.

We also applied the QCM method for *E. coli* detection at the surface of piezo crystal modified by DNA aptamers. For thiolated aptamers, we observed that the surface binding was rather weak, and it was necessary to lower the pH leading to about a 10 Hz decrease in the resonant frequency in the flow following the addition of thiolated aptamer (Figure 5). The resulting change in frequency after the addition of around 10^4 CFU/mL was hard to distinguish from the noise and drift of the QCM system. Viscoelastic analysis using the Voinova–Voigt viscoelastic model [6] shows that the estimated height of this aptamer is about 5 nm at layer formation and that the resulting height is below 1 nm (Figure 6). Equations used for Voinova model are presented below:

$$\Delta f = -\frac{1}{2\pi\rho_q h_q}\left(\frac{\eta_l}{\delta} + h_f\rho_f\omega - 2h_f\left(\frac{\eta_l}{\delta}\right)^2\frac{\eta_f\omega^2}{\mu_f^2 + \omega^2\eta_f^2}\right)$$

$$\Delta D = \frac{1}{2\pi f_0\rho_q h_q}\left(\frac{\eta_l}{\delta} + h_f\rho_f\omega + 2h_f\left(\frac{\eta_l}{\delta}\right)^2\frac{u_f\omega}{\mu_f^2 + \omega^2\eta_f^2}\right)$$

where $\omega = 2\pi f_0 n$, $\delta = \sqrt{\frac{2\eta_l}{\eta_l\omega}}$, q, l, and f subscripts are designations for quartz, liquid, and film, respectively.

The length of unfolded aptamer is around 15 nm, indicating that the aptamer at the crystal surface is likely to be in a different conformation than the denatured random. However, low altitude can also imply insufficient coverage (as it is an average height) or an aptamer lying on the surface. These results are also supported by the change in viscosity and elasticity. Shearing modulus, μ, increased after leaching from the layer, implying that the aptamer had a greater tendency to return, implying closer association with the layer. On the contrary, the viscosity decreased, which may mean that the aptamer is not in a tense conformation and, thus, lies on the surface of the gold electrode. To circumvent this problem, it is probably necessary to modify the layer with small thiol molecules in sufficient proportion to the aptamer so that the aptamer is in a more advantageous position and has room to bind bacteria. The optical method based on AuNPs had the highest estimated LOD. The light reflectometry method was able to measure *E. coli* in the range of 10^3 to 10^4 CFU/mL; however, sensitivity as low as 100 CFU/mL can also be attained [7]. The acoustic method provides advantages in the possibility of probing the viscoelastic changes in the surfaces; however, the bacteria can negatively or positively induce the changes in the frequency, which proves a challenge when it is used as a biosensor [8,9].

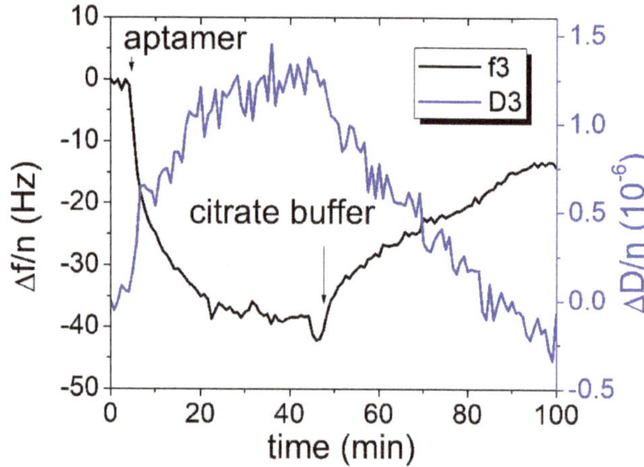

Figure 5. The changes in frequency ($\Delta f/n$) and dissipation ($\Delta D/n$) following addition of thiolated aptamer in a concentration of 5 uM and washing the surface by citrate buffer. The results were obtained from TSM experiment for 3rd overtone. The changes in frequency (black) and dissipation (blue) are divided by the overtone number n=3. The moment of addition of aptamers and buffer are shown by arrows.

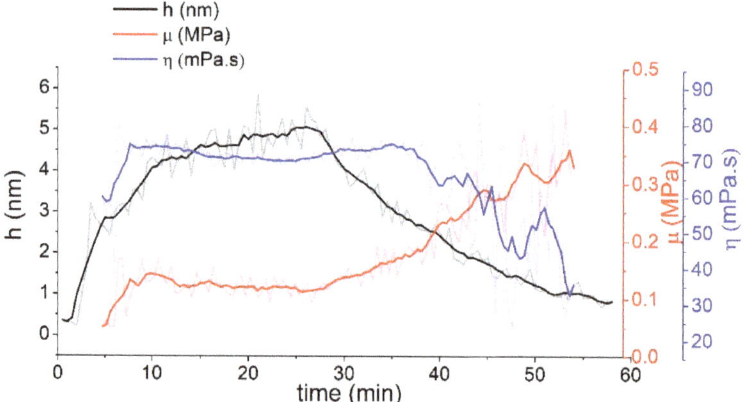

Figure 6. The kinetic of changes in the thickness (h—black), viscosity (η, red), and shear modulus (μ, blue) based on the Voinova–Voigt viscoelastic model using the data on the changes in resonant frequency and dissipation obtained from TSM experiment.

Author Contributions: Conceptualization, T.H. and S.E.K.; methodology, P.P.; formal analysis, I.P.; investigation, I.P., J.S., M.A. and P.P.; resources, L.B.; writing—original draft preparation, I.P.; writing—review and editing, I.P. and T.H.; visualization, I.P.; supervision, S.E.K., P.P. and T.H.; project administration, T.H.; funding acquisition, S.E.K. and T.H. All authors have read and agreed to the published version of the manuscript.

Funding: This project has received funding from the European Union's Horizon 2020 research and innovation programme under the Marie Skłodowska-Curie (grant agreement no. 101007299) and from Science Agency VEGA (project no. 1/0419/20) (to T.H.).

Institutional Review Board Statement: Not applicable.

Informed Consent Statement: Not applicable.

Data Availability Statement: The data presented in this study are available on request from the corresponding author.

Conflicts of Interest: The authors declare no conflict of interest.

References

1. Hoagland, L.; Ximenes, E.; Ku, S.; Ladisch, M. Foodborne pathogens in horticultural production systems: Ecology and mitigation. *Sci. Hortic.* **2018**, *236*, 192–206. [CrossRef]
2. Priyanka, B.; Patil, R.K.; Dwarakanath, S. A review on detection methods used for foodborne pathogens. *Indian J. Med. Res.* **2016**, *144*, 327. [CrossRef] [PubMed]
3. Safavieh, M.; Nahar, S.; Zourob, M.; Ahmed, M.U. Microfluidic biosensors for high throughput screening of pathogens in food. In *High Throughput Screening for Food Safety Assessment*; Elsevier: Amsterdam, The Netherlands, 2015; pp. 327–357. [CrossRef]
4. Siddiqui, S.; Yuan, J. Binding Characteristics Study of DNA based Aptamers for *E. coli* O157:H7. *Molecules* **2021**, *90*, 204. [CrossRef] [PubMed]
5. Lyu, Y.; Teng, I.T.; Zhang, L.; Guo, Y.; Cai, R.; Zhang, X.; Qiu, L.; Tan, W. Comprehensive Regression Model for Dissociation Equilibria of Cell-Specific Aptamers. *Anal. Chem.* **2018**, *90*, 10487–10493. [CrossRef] [PubMed]
6. Voinova, M.V. The theory of acoustic sensors application in air quality control. *Urban Clim.* **2018**, *24*, 264–275. [CrossRef]
7. Angelopoulou, M.; Tzialla, K.; Voulgari, A.; Dikeoulia, M.; Raptis, I.; Kakabakos, S.E.; Petrou, P. Rapid Detection of Salmonella typhimurium in Drinking Water by a White Light Reflectance Spectroscopy Immunosensor. *Sensors* **2021**, *21*, 2683. [CrossRef] [PubMed]
8. Alexander, T.E.; Lozeau, L.D.; Camesano, T.A. QCM-D characterization of time-dependence of bacterial adhesion. *Cell Surf.* **2019**, *5*, 100024. [CrossRef] [PubMed]
9. Tatarko, M.; Spagnolo, S.; Oravczová, V.; Süle, J.; Hun, M.; Hucker, A.; Hianik, T. Changes of viscoelastic properties of aptamer based sensing layers following interaction with Listeria innocua. *Sensors* **2021**, *21*, 5585. [CrossRef] [PubMed]

engineering proceedings

Proceeding Paper

Detection of Prostate Cancer Biomarker PCA3 with Electrochemical Apta-Sensor [†]

Sarra Takita [1],* , Alexei Nabok [1], Anna Lishchuk [2], Magdi H. Mussa [1,3] and David Smith [4]

1. Material and Engineering Research Institute (MERI), Sheffield Hallam University, Sheffield S1 1WB, UK; engan@exchange.shu.ac.uk (A.N.); magdimosa1976@gmail.com (M.H.M.)
2. Department of Chemistry, University of Sheffield, Brook Hill, Sheffield S3 7HF, UK; a.lishchuk@sheffield.ac.uk
3. The Institute of Marine Engineering, Science and Technology, London SW1H 9JJ, UK
4. Biomolecular Sciences Research Centre (BMRC), Sheffield Hallam University, Sheffield S1 1WB, UK; hwbds1@exchange.shu.ac.uk
* Correspondence: sarah.a.takita@gmail.com; Tel.: +44-740731366
† Presented at the 2nd International Electronic Conference on Biosensors, 14–18 February 2022; Available Online: https://sciforum.net/event/IECB2022.

Abstract: This is a continuation of our research into the development of novel biosensing technologies for early diagnostics of prostate cancer (PCa). The existing PCa diagnostics based on PSA detection (prostate cancer antigen) in blood serum often yield controversial outcomes and require improvement. At the same time, the long non-coded RNA transcript PCA3 overexpressed in PCa patients' urine proved to be an ideal biomarker for PCa diagnosis, and recent research mainly focuses on developing biosensors for the detection of PCA3. One of the most promising directions in this research is the use of aptamers as bio-receptors for PCA3. We demonstrated the earlier great potential of electrochemical sensors exploiting aptamer labelled with redox group ferrocene. In this work, we use the RNA-based aptamer specific to 227 nt fragment of lncRNA PCA3 labelled with methylene blue redox label which offers a higher affinity to PCA3 than commonly used DNA-based aptamers. Before proceeding with biosensing experiments, the gold screen-printed electrodes were cleaned by CV scanning in a sulfuric acid solution, which removed surface contaminations and thus improved immobilization of aptamers. The quality of the gold surface was assessed by contact angle measurements. Moreover, the concentration of immobilized aptamers was optimized to achieve the best results in electrochemical measurements. Initial tests were carried out using cyclic voltammograms (CV) measurements and showed a correlation between oxidation/reductions peaks intensities and the concentration of PCA3. Such experiments proved the main concept of the proposed apta-sensing, e.g., the changes of aptamer secondary structure during binding the target (PCA3) resulting in redox labels coming closer to the electrode surface and thus increasing the charge transfer. The lowest recorded concentration of PCA3 was 0.01 nM in CV measurements, which is close to the LDL level for this method. Much more promising results were obtained with the electrochemical impedance spectroscopy (EIS) measurements, which showed remarkable features of increasing sensitivity at low concentrations of PCA3. The extrapolation of data below 0.05 nM level allowed estimating LDL of about 0.4 pM. The results obtained are very encouraging and constitute a major step towards developing a simple, reliable, and cost-effective diagnostic tool for the early detection of prostate cancer.

Keywords: prostate cancer; lncRNA PCA3 biomarker; RNA-based aptamer; electrochemical biosensor; gold screen-printed electrodes; cyclic voltammograms; electrochemical impedance spectroscopy

Citation: Takita, S.; Nabok, A.; Lishchuk, A.; Mussa, M.H.; Smith, D. Detection of Prostate Cancer Biomarker PCA3 with Electrochemical Apta-Sensor. *Eng. Proc.* **2022**, *16*, 8. https://doi.org/10.3390/IECB2022-12257

Academic Editors: Giovanna Marrazza and Sara Tombelli

Published: 14 February 2022

Publisher's Note: MDPI stays neutral with regard to jurisdictional claims in published maps and institutional affiliations.

Copyright: © 2022 by the authors. Licensee MDPI, Basel, Switzerland. This article is an open access article distributed under the terms and conditions of the Creative Commons Attribution (CC BY) license (https://creativecommons.org/licenses/by/4.0/).

1. Introduction

Prostate cancer (PCa), also known as adenocarcinoma, is the most common worldwide type of cancer in men after lung cancer, and it is the second leading cause of mortality among men [1,2]. Clinically, the standard diagnostics test for detection of PCa is based on the detection and quantification of total serum prostate-specific antigen (tPSA) in serum, followed by further examinations such as digital rectal examination (DRE) and imaging investigations if PCa is suspected [3,4]. Despite the benefits of these tests, physicians still have difficulty identifying early-stage PCa due to its asymptomatic nature and/or symptoms resembling benign conditions, such as prostatic hyperplasia (BPH) [5]. Furthermore, PSA testing has limits in terms of specificity, accuracy, and sensitivity [6–8]. Hence, identifying other specific PCa biomarkers besides PSA for the detection of PCa is of high importance [9,10]. The long non-coding RNA (lncRNA) known as PCA3 overexpressed in PCa patients' urine has been widely accepted as one of the specific biomarkers for malignant prostate cells [11–13]. PCA3 level can predict prostatic biopsies' outcome, especially in combination with other PCa biomarkers such as PSA and can reduce the likelihood of false-positive results [14–16]. Prognesa® test, approved in USA in 2012, is based on detection of both PCA3 and PSA using quantitative nucleic acid amplification after digital rectal examination and yields a PCA3 score (the ratio of PCA3 to PSA mRNA molecules in urine specimens) [17,18]. However, such a test is time-consuming, expensive, and requires highly skilled operators. Biosensing techniques involving aptamers as bioreceptors, e.g., single-stranded RNA or DNA molecules with high affinity for target molecules, are alternatives to well-established immunosensors. A GC3 RNA aptamer against the 277 nt section of lncRNA transcript PCA3 was developed using SELEX process and reported by Marangoni et al. [19]. According to this study, the GC3 aptamer showed the highest affinity towards PCA3. From the transducer point of view, electrochemical sensors are the most attractive because of their high sensitivity, simplicity of operation and low cost [20–22]. The concept of detection of PCA3 using specific DNA-based aptamers labelled with redox group (ferrocene) was explained and proved in our recent publications [23,24].

This work is a further study of the implementation of electrochemical sensing combined with redox-labelled aptamers for the detection of PCA3. Here, we used the original GC3 RNA-based aptamer, which is supposed to provide higher specificity towards PCA3 target. It is also labelled with another redox chemical, e.g., methylene blue. Two electrochemical methods of cyclic voltammetry (CV) and electrochemical impedance spectroscopy (EIS) were exploited here using screen-printed gold electrodes (SPGE) and interdigitated gold electrodes (IDGE), respectively, and compared the sensitivity of detection with so-far published papers. The properties of the gold surface of these electrodes and its effect on electrochemical measurements were also assessed in this study. CV is a common electrochemical method for analysing redox reactions at the electrode surface [25], while EIS is a very sensitive method capable of detecting tiny changes in a double layer on the electrode surface [26]. The high sensitivity of EIS could be beneficial for apta-sensing and for the detection of PCA3, as has been shown in [24]. The results obtained are very encouraging and constitute a major step towards developing a simple, reliable, and cost-effective diagnostic tool for the early detection of prostate cancer.

2. Materials and Methods

2.1. Chemicals and Reagents

The CG RNA-based aptamer reproducing the following sequence of nucleotides 5′-AGUUUUUGCGUGUGCCCUUUUUGUCCCC-3′ published by the inventors [19] was customised by Merck Life Science Ltd. (Dorset, UK). This aptamer has been used in our previous work [23]. The methylene blue and thiol groups were attached to C5 and C3 termini, respectively. The target analyte, e.g., the 277 nt fragment of lncRNA PCA3 was purchased from Eurofins Scientific (Guildford, UK branch) and prepared in PBS (pH 7.5).

HEPES binding buffer (HBB) pH 7.6, sodium phosphate di-basic (Na_2HPO_4), potassium phosphate mono-basic (KH_2PO_4), potassium chloride (KCl), magnesium chloride ($MgCl_2$), dithiothreitol (DTT), and sodium chloride (NaCl), were procured from Sigma-Aldrich (Gillingham, UK). All reagents were of analytical grade. The target analyte, e.g., the 277 nt fragment of lncRNA PCA3 was purchased from Eurofins Scientific (Guildford, UK) and prepared in PBS (pH 7.5). All aqueous solutions were prepared using 18.2 MΩ·cm deionized (DI) water (Millipore, Watford, UK). The methylene blue labelled CG RNA based aptamer [19], which was used in our previous work [23], was acquired from Merck Life Science Ltd. (London branch, UK). The methylene blue and thiol groups were attached to C5 and C3 termini, respectively.

2.2. Measurements and Instrumentation

Three-electrode gold screen-printed assemblies (AT) with Ag/AgCl reference electrodes with 4 mm diameter working electrodes from DropSens Metrohm (Runcorn, UK) were used for CV measurements with Dropsens potentiostat Stat8000. Voltage ranges from -0.4 to 0.2 V with the step of 10 mV and scan rate of 40 mV/s were used. CV cycles were recorded 5 times until the current readings were stabilized. In addition to cyclic voltammetry, the time dependencies of cathodic current at -0.25 V were recorded on electrodes during exposure to PCA3 of different concentrations for kinetic study of the PCA3 to aptamer binding. Interdigitated gold electrodes having 50 fringes with the spacing of 5 μm were used for EIS measurements with Parastate 4000 impedance analyzer. The AC signal of 50 mV amplitude (and zero DC bias) with the frequency varied from 0.1 Hz to 100 kHz was used in these measurements. Moreover, the sessile drop method was used in water contact angle measurements with OCA 15EC Goniometer. DI water 10 μL drops were dispensed on top of the electrode surface, and the droplet microscopic images were captured and analysed using built-in software.

2.3. Immobilization of Aptamers and Preparation for CV and EIS Measurements

The aptamers were immobilized on the surface of both types of electrodes (SPGE and IDGE) following the procedure described in detail in our earlier publications [24,27]. Extra cleaning of the electrodes method has been done using CV cleaning scans in 0.1 M H_2SO_4 until the gold oxide reduction peak no longer increased in size. Such treatment removes surface contaminants without damaging the gold surface as described in [20].

This additional cleaning procedure has resulted in consistent and a smooth electrode surface. The target analyte (PCA3) was resuspended in detection buffer (HEPES pH 7.5, 120 mm NaCl, 5 mm KCl,) at different concentrations from 100 nM down to 0.01 nM and were used in both CV and EIS measurements.

3. Results and Discussion

3.1. Characterization of Gold Electrodes after Cleaning

Characterization of gold surface wettability before and after CV cleaning in 0.1 mM H_2SO_4 solution was carried out by contact angle measurement. Figure 1a shows a typical CV of SPGE during cleaning, similar to that described in [20]. Figure 1b shows the effect of CV cleaning on the wettability of various gold electrodes surfaces using the contact angle measurement data. Images of water droplets are shown below.

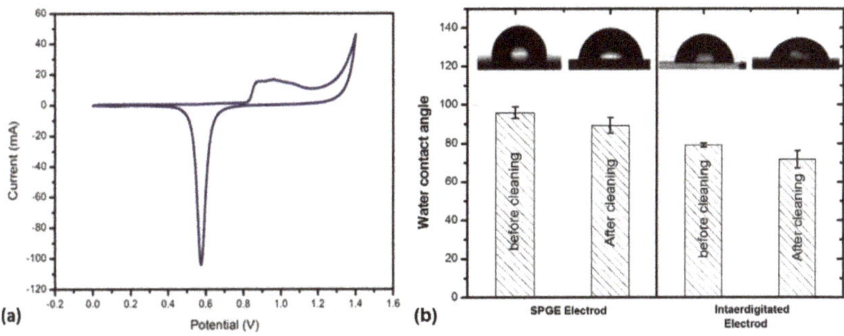

Figure 1. Cleaning and characterization of gold surfaces: (**a**) Typical CV curve of SPGE in 0.1 M H_2SO_4 solution cleaning responses. (**b**) Results of contact angle measurements of SPGE and IDGE. Images of water droplets are shown above.

Since the cleaning methodology is essential, as is described in [28]. A minimum of three sets of measurements were performed across the surface of each sample. The homogeneity and structure of the gold electrode surface influence the peak currents in CV and amperometry detection [29]. For this reason, the gold electrode surface must be cleaned before the immobilization of aptamers. The final stabilized CV curve after 10 cycles of cleaning is shown in Figure 1a. The reduction in contact angles for both electrodes AT SPGE and IDGE after cleaning indicates better wettability of the gold surface is shown in Figure 1b.

3.2. Electrochemical Apta-Sensing of PCA3

Typical CV curves were recorded on SPGE with immobilized aptamers before and after exposure to PCA3 of different concentrations, as shown in Figure 2. The characteristic oxidation and reduction peaks of methylene blue appeared on all CV curves at around −0.2 and −0.25 V, respectively. The intensity of redox current peaks varied dramatically depending on the concentration of PCA3 bound to aptamers. Initial peaks for aptamers appear as small humps on CV curves. The intensity of these peaks increases progressively with the increase in PCA3 concentration. This trend can be observed clearly on the inset in Figure 2, showing concentration dependence of the absolute values of changes in cathodic current at −0.25 V. The values of |ΔIc| were calculated by subtracting the baseline Ic value for pure aptamer from Ic values corresponding to different concentrations of PCA3.

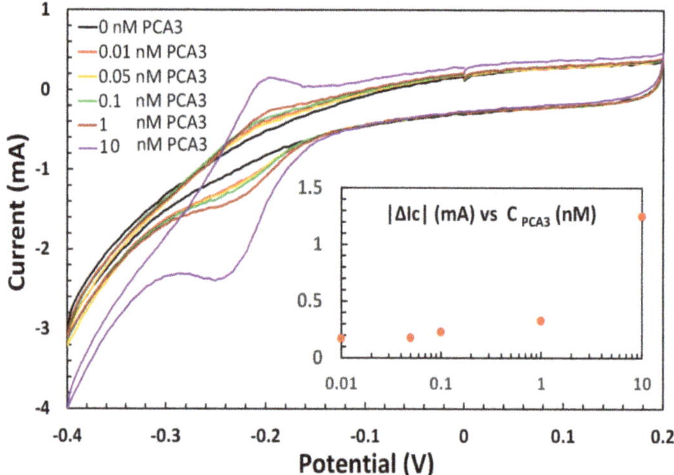

Figure 2. Typical CV curves recorded on SPGE functionalized with aptamers before and after exposure to PCA3 of different concentrations. Inset shows the dependence of absolute values of changes in cathodic current at −0.25 V on PCA3 concentration.

The graph in Figure 2 (inset) represents the beginning of a standard sigmoid curve. The saturation of the sensor response could be achieved at much larger concentrations of PCA3. At low concentrations of PCA3 the response is almost flattened between 0.05 and 0.01 nM, which means that the LDL can be estimated as 0.05 nM. This value corresponds to 0.125 ng/mL, considering the molecular weight of about 2.5 kD for a 78 bp fragment of lncRNA PCA3), which is close to LDL = 0.1 ng/mL evaluated previously [24] for CV measurements for ferrocene-labelled aptamer. It should be noted that the lowest concentration of PCA3 detected on SPGE before CV cleaning was only 0.1 nM. Therefore, electrodes' cleaning resulted in 5 to 10 folds increase in the sensitivity. The optimal concentration of aptamers yielding less noisy CV graphs was 1 μM.

3.3. Electrochemical Impedance Spectroscopy (EIS)

The results of EIS measurements are presented in Figure 3 as the dependencies of the imaginary (Z_{im}) vs real (Z_{re}) parts of impedance known as Nyquist plots. As one can see, all Nyquist plots recorded on IDGE with immobilized aptamers before and after exposure to different concentrations of PCA3 appear as almost ideal semi-circles with the anticlockwise direction of AC frequency increase. This is a clear indication of the negligible contribution of diffusion of redox chemicals to the electrode surface, which is obvious since the redox labels are attached to aptamers. Therefore, the behaviour of IDGE modified with redox-labelled aptamers can be described by a simplified (without diffusion impedance) equivalent circuit shown as an inset in Figure 3a. The sizes (diameters) of Nyquist semi-circles correlate with the concentration of PCA3 bound to the aptamers. The layer of fresh aptamers shows the largest diameter, which decreases with an increase in PCA3 concentration.

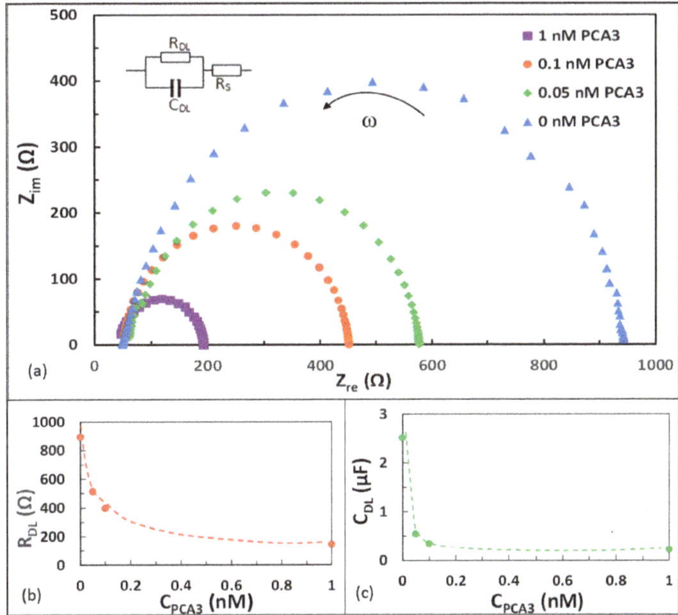

Figure 3. (a) Nyquist graphs recorded on IDGE modified with aptamers before and after exposure to different concentrations of PCA3. Arrow indicates the direction of frequency increase. A simplified equivalent circuit is shown as inset. Dependencies of the resistance (b) and capacitance (c) of a double layer on the concentration PCA3.

Analysis of Nyquist graphs is based on the formula of electrical impedance (Z) of a simplified equivalent circuit (see inset in Figure 3a) described earlier [24]:

$$Z = Z_{re} - jZ_{im}; \quad Z_{re} = \frac{R_{DL}}{1 + \omega^2 R_{DL}^2 C_{DL}^2} + R_S; \quad Z_{im} = \frac{\omega R_{DL}^2 C_{DL}}{1 + \omega^2 R_{DL}^2 C_{DL}^2},$$

where Z_{re} and Z_{im} are respectively the real and imaginary parts of impedance, which depend on the resistance and capacitance of a double layer (R_{DL} and C_{DL}) and bulk resistance of the solution (R_S) as well as the frequency (ω) of AC signal. Analysis of the above equation gives at low frequencies ($\omega \to 0$) $Z_{re}^0 = R_{DL} + R_S$ and $Z_{im}^0 = 0$, while at high frequencies ($\omega \to \infty$) $Z_{re}^\infty = R_S$ and $Z_{im}^\infty = 0$. Therefore, the characteristic parameter of double-layer resistance R_{DL} can be calculated as $R_{DL} = R_{re}^0 - Z_{Re}^\infty$.

On the other hand, the capacitance of a double layer (C_{DL}) can be calculated from the maximal values of $Z_{im}^{max} \approx 1/\omega C_{DL}$.

The values of R_{DL} and C_{DL} were calculated from the Nyquist plots in Figure 3a for all concentrations of PCA3, including zero concentration corresponding to the aptamer layer before binding PCA3, and the results are given in Figure 3b,c. As one can see from Figure 3b,c, the sensitivity of detection increases with the decrease in PCA3 concentration, which is opposite to CV measurements (see inset in Figure 2). Unfortunately, it is impossible to precisely evaluate the LDL of EIS measurements because the lowest concentration of PCA3 was only 0.05 nM. However, it is possible to estimate the LDL from the slope of the R_{DL} vs C_{PCA3} graph at low concentrations (between 0.05 and 0 nM) in Figure 3b. The gradient of this graph $\Delta C_{PCA3}/\Delta R_{DL} = 0.05/378.7 = 1.3 \times 10^{-4}$ nM/Ω, therefore assuming the triple noise level of about 3 Ω, the LDL can be estimated as 0.4 pM. A similar estimation of LDL can be done from the C_{DL} vs C_{PCA3} graph in Figure 3c. The gradient $\Delta C_{PCA3}/\Delta C_{DL} = 0.05/2.5 = 0.02$ nM/μF. Assuming the triple noise level of 0.01 μF, the LDL of 0.2 pM can be estimated.

4. Conclusions

The results obtained in this work proved a concept of electrochemical detection of prostate cancer marker PCA3 using RNA-based aptamers labelled with methylene blue redox group. The mechanism of detection is based on the increasing of charge transfer between the redox label and the electrode because of aptamers engulfing the target molecules (PCA3) and bringing redox labels (methylene blue) closer to the electrode surface. One of the main advantages of such an approach is the absence of redox chemicals in solution, which may allow performing tests on real samples of urine in future. Cyclic voltammogram measurements resulted in moderate LDL values between 0.05 and 0.01 nM, similar to the values obtained in our previous work on DNA-based aptamers labelled with ferrocene [25]. Such sensitivity should be sufficient for clinical use. However, EIS method appeared to be much more promising because of a different nature of the sensor response having the sensitivity increasing at low concentrations. The absence of redox chemicals in the solution allowed using a simplified model for EIS data analysis, which resulted in a correlation between both the resistance and capacitance of a double layer on the electrode surface and the concentration of PCA3 consistent with the above model of aptasensing. Our estimations showed that LDL for EIS measurements could be in sub-pM range. Additionally, some important technological steps of sensors preparation, such as electrochemical cleaning of gold screen-printed electrodes and optimization of the aptamer concentration for immobilization on the surface of gold, were implemented. Overall, this work is a major step towards the future development of a novel methodology of prostate cancer diagnostics. This work is ongoing and focuses on more detailed CV and EIS measurements in a wide range of PCA3 concentrations, statistical analysis of data, and more precise evaluation of LDL. PCA3 detection in complex media such as urine will be attempted.

Author Contributions: All authors contribute to the manuscript equally. All authors have read and agreed to the published version of the manuscript.

Funding: This research received no external funding.

Institutional Review Board Statement: Not applicable.

Informed Consent Statement: Not applicable.

Data Availability Statement: The data are not publicly available; The data files are stored on corresponding instruments and on personal computers.

Acknowledgments: We would like to acknowledge Sheffield Hallam University, UK, specifically material and engineering research institute (MERI) and, Biomolecular Sciences Re-search Centre (BMRC), in City Campus, Howard Street, Sheffield, S1 1WB, for full access to its resources and material in this research.

Conflicts of Interest: The authors declare no conflict of interest.

References

1. Ferlay, J.; Soerjomataram, I.; Dikshit, R.; Eser, S.; Mathers, C.; Rebelo, M.; Parkin, D.M.; Forman, D.; Bray, F. Cancer incidence and mortality worldwide: Sources, methods and major patterns in GLOBOCAN 2012. *Int. J. Cancer* **2015**, *136*, E359–E386. [CrossRef] [PubMed]
2. Sung, H.; Ferlay, J.; Siegel, R.L.; Laversanne, M.; Soerjomataram, I.; Jemal, A.; Bray, F. Global Cancer Statistics 2020: GLOBOCAN Estimates of Incidence and Mortality Worldwide for 36 Cancers in 185 Countries. *CA Cancer J. Clin.* **2021**, *71*, 209–249. [CrossRef]
3. Salman, J.W.; Schoots, I.G.; Carlsson, S.V.; Jenster, G.; Roobol, M.J. Prostate Specific Antigen as a Tumor Marker in Prostate Cancer: Biochemical and Clinical Aspects. In *Advances in Cancer Biomarkers*; Springer: Dordrecht, The Netherlands, 2015; Volume 867, pp. 93–114, ISBN 9789401772150.
4. Buzzoni, C.; Auvinen, A.; Roobol, M.J.; Carlsson, S.; Moss, S.M.; Puliti, D.; de Koning, H.J.; Bangma, C.H.; Denis, L.J.; Kwiatkowski, M.; et al. Metastatic Prostate Cancer Incidence and Prostate-specific Antigen Testing: New Insights from the European Randomized Study of Screening for Prostate Cancer. *Eur. Urol.* **2015**, *68*, 885–890. [CrossRef]
5. Daniyal, M.; Siddiqui, Z.A.; Akram, M.; Asif, H.M. MINI-REVIEW Epidemiology. *Etiol. Diagn. Treat. Prostate Cancer* **2014**, *15*, 9575–9578.
6. Heijnsdijk, E.A.M.; der Kinderen, A.; Wever, E.M.; Draisma, G.; Roobol, M.J.; de Koning, H.J. Overdetection, overtreatment and costs in prostate-specific antigen screening for prostate cancer. *Br. J. Cancer* **2009**, *101*, 1833–1838. [CrossRef] [PubMed]
7. Aslan, G.; Irer, B.; Cimen, S.; Goktay, Y.; Celebi, I.; Tuna, B.; Yorukoglu, K. The Performance of Abnormal Digital Rectal Examination for the Detection of Prostate Cancer at Stratified Prostate Specific Antigen Levels. *Open J. Urol.* **2011**, *1*, 67–71. [CrossRef]
8. Hussain, S.; Gunnell, D.; Donovan, J.; McPhail, S.; Hamdy, F.; Neal, D.; Albertsen, P.; Verne, J.; Stephens, P.; Trotter, C.; et al. Secular trends in prostate cancer mortality, incidence and treatment: England and Wales, 1975–2004. *BJU Int.* **2008**, *101*, 547–555. [CrossRef]
9. Mistry, K.; Cable, G. Meta-analysis of prostate-specific antigen and digital rectal examination as screening tests for prostate carcinoma. *J. Am. Board Fam. Pract.* **2003**, *16*, 95–101. [CrossRef] [PubMed]
10. Altuwaijri, S. Role of Prostate Specific Antigen (PSA) in Pathogenesis of Prostate Cancer. *J. Cancer Ther.* **2012**, *3*, 331–336. [CrossRef]
11. Chistiakov, D.A.; Myasoedova, V.A.; Grechko, A.V.; Melnichenko, A.A.; Orekhov, A.N. New biomarkers for diagnosis and prognosis of localized prostate cancer. *Semin. Cancer Biol.* **2018**, *52*, 9–16. [CrossRef]
12. Rönnau, C.G.H.; Verhaegh, G.W.; Luna-Velez, M.V.; Schalken, J.A. Noncoding RNAs as Novel Biomarkers in Prostate Cancer. *BioMed Res. Int.* **2014**, *2014*, 591703. [CrossRef] [PubMed]
13. Schalken, J.A.; Hessels, D.; Verhaegh, G. New targets for therapy in prostate cancer: Differential display code 3 (DD3PCA3), a highly prostate cancer-specific gene. *Urology* **2003**, *62*, 34–43. [CrossRef]
14. Bourdoumis, A.; Papatsoris, A.G.; Chrisofos, M.; Efstathiou, E.; Skolarikos, A.; Deliveliotis, C. The novel prostate cancer antigen 3 (PCA3) biomarker. *Int. Braz. J. Urol.* **2010**, *36*, 665–669. [CrossRef]
15. Wu, A.K.; Reese, A.C.; Cooperberg, M.R.; Sadetsky, N.; Shinohara, K. Utility of PCA3 in patients undergoing repeat biopsy for prostate cancer. *Prostate Cancer Prostatic Dis.* **2012**, *15*, 100–105. [CrossRef] [PubMed]
16. Goode, R.R.; Marshall, S.J.; Duff, M.; Chevli, E.; Chevli, K.K. Use of PCA3 in detecting prostate cancer in initial and repeat prostate biopsy patients. *Prostate* **2013**, *73*, 48–53. [CrossRef] [PubMed]
17. Groskopf, J.; Aubin, S.M.; Deras, I.L.; Blase, A.; Bodrug, S.; Clark, C.; Brentano, S.; Mathis, J.; Pham, J.; Meyer, T.; et al. APTIMA PCA3 Molecular Urine Test: Development of a Method to Aid in the Diagnosis of Prostate Cancer. *Clin. Chem.* **2006**, *52*, 1089–1095. [CrossRef] [PubMed]
18. Deras, I.L.; Aubin, S.M.J.; Blase, A.; Day, J.R.; Koo, S.; Partin, A.W.; Ellis, W.J.; Marks, L.S.; Fradet, Y.; Rittenhouse, H.; et al. PCA3: A Molecular Urine Assay for Predicting Prostate Biopsy Outcome. *J. Urol.* **2008**, *179*, 1587–1592. [CrossRef] [PubMed]
19. Marangoni, K.; Neves, A.F.; Rocha, R.M.; Faria, P.R.; Alves, P.T.; Souza, A.G.; Fujimura, P.T.; Santos, F.A.A.; Araújo, T.G.; Ward, L.S.; et al. Prostate-specific RNA aptamer: Promising nucleic acid antibody-like cancer detection. *Sci. Rep.* **2015**, *5*, 12090. [CrossRef]
20. Butterworth, A.; Blues, E.; Williamson, P.; Cardona, M.; Gray, L.; Corrigan, D.K. SAM Composition and Electrode Roughness Affect Performance of a DNA Biosensor for Antibiotic Resistance. *Biosensors* **2019**, *9*, 22. [CrossRef] [PubMed]
21. Fang, X.; Jin, Q.; Jing, F.; Zhang, H.; Zhang, F.; Mao, H.; Xu, B.; Zhao, J. Integrated biochip for label-free and real-time detection of DNA amplification by contactless impedance measurements based on interdigitated electrodes. *Biosens. Bioelectron.* **2013**, *44*, 241–247. [CrossRef]
22. Uludag, Y.; Narter, F.; Sağlam, E.; Köktürk, G.; Gök, M.Y.; Akgün, M.; Barut, S.; Budak, S. An integrated lab-on-a-chip-based electrochemical biosensor for rapid and sensitive detection of cancer biomarkers. *Anal. Bioanal. Chem.* **2016**, *408*, 7775–7783. [CrossRef] [PubMed]
23. Takita, S.; Nabok, A.; Lishchuk, A.; Smith, D. Optimization of Apta-Sensing Platform for Detection of Prostate Cancer Marker PCA3. *Int. J. Mol. Sci.* **2021**, *22*, 12701. [CrossRef]

24. Nabok, A.; Abu-Ali, H.; Takita, S.; Smith, D.P. Electrochemical detection of prostate cancer biomarker pca3 using specific rna-based aptamer labelled with ferrocene. *Chemosensors* **2021**, *9*, 59. [CrossRef]
25. Elgrishi, N.; Rountree, K.J.; McCarthy, B.D.; Rountree, E.S.; Eisenhart, T.T.; Dempsey, J.L. A Practical Beginner's Guide to Cyclic Voltammetry. *J. Chem. Educ.* **2018**, *95*, 197–206. [CrossRef]
26. Bard, A.J.; Faulkner, L.R. *Electrochemical Methods: Fundamentals and Applications*, 2nd ed.; John Wiley & Sons Inc.: New York, NY, USA, 2001; ISBN 978-0-471-04372-0.
27. Takita, S.; Nabok, A.; Smith, D.; Lishchuk, A. Spectroscopic Ellipsometry Detection of Prostate Cancer Bio-Marker PCA3 Using Specific Non-Labeled Aptamer: Comparison with Electrochemical Detection. *Chem. Proc.* **2021**, *5*, 65. [CrossRef]
28. Mussa, M.H.; Farmilo, N.; Lewis, O. The Influence of Sample Preparation Techniques on Aluminium Alloy AA2024-T3 Substrates for Sol-Gel Coating. *Eng. Proc.* **2021**, *11*, 5. [CrossRef]
29. Fischer, L.M.; Tenje, M.; Heiskanen, A.R.; Masuda, N.; Castillo, J.; Bentien, A.; Émneus, J.; Jakobsen, M.H.; Boisen, A. Gold cleaning methods for electrochemical detection applications. *Microelectron. Eng.* **2009**, *86*, 1282–1285. [CrossRef]

Proceeding Paper

A Portable Screening Device for SARS-CoV-2 with Smartphone Readout †

Antonios Georgas *, Konstantinos Agiannis, Vasiliki Papakosta, Spyridon Angelopoulos, Angelo Ferraro and Evangelos Hristoforou

School of Electrical and Computer Engineering, National Technical University of Athens, 15780 Athens, Greece; k.agiannis@yandex.com (K.A.); papakostavasiliki@gmail.com (V.P.); spyrosag@central.ntua.gr (S.A.); an.ferraro2@gmail.com (A.F.); hristoforou@ece.ntua.gr (E.H.)
* Correspondence: antonisgeo@mail.ntua.gr
† Presented at the 2nd International Electronic Conference on Biosensors, 14–18 February 2022; Available Online: https://sciforum.net/event/IECB2022.

Abstract: Since the outbreak of the COVID-19 pandemic, great emphasis has been placed on the development of rapid virus detection devices, the principle of operation of many of which is the detection of the virus structural protein spike. Although several such devices have been developed, most are based on the visual observation of the result, without providing the possibility of its electrical processing. This paper presents a biosensor platform for the rapid detection of spike proteinboth in laboratory conditions and in swab samples from hospitalized patients. The platform consists of a microcontroller-based readout circuit, which measures the capacitance change generated in an interdigitated electrode transducer by the presence of the spike protein. The circuit efficiency is calibrated by its correlation with the capacitance measurement of an LCR meter. The test result is made available in less than 2 min through the microcontroller's LCD screen, and at the same time, the collected data are sent wirelessly to a mobile application interface. In this way, the continuous and effective screening of SARS-CoV-2 patients is facilitated and enhanced, providing big data statistics of COVID-19 in terms of space and time.

Keywords: biosensor; SARS-CoV-2; spike; screening; readout; smartphone; portable; interdigitated electrodes

check for **updates**

Citation: Georgas, A.; Agiannis, K.; Papakosta, V.; Angelopoulos, S.; Ferraro, A.; Hristoforou, E. A Portable Screening Device for SARS-CoV-2 with Smartphone Readout. *Eng. Proc.* **2022**, *16*, 7. https://doi.org/10.3390/IECB2022-12274

Academic Editors: Giovanna Marrazza and Sara Tombelli

Published: 14 February 2022

Publisher's Note: MDPI stays neutral with regard to jurisdictional claims in published maps and institutional affiliations.

Copyright: © 2022 by the authors. Licensee MDPI, Basel, Switzerland. This article is an open access article distributed under the terms and conditions of the Creative Commons Attribution (CC BY) license (https://creativecommons.org/licenses/by/4.0/).

1. Introduction

The COVID-19 pandemic has proven to be a major threat to humanity. Scientists all over the world are fighting against it, which includes developing new technologies for the detection of SARS-CoV-2.

The standard method for virus detection is real-time PCR. Despite its high efficiency, real-time PCR is a time-consuming and costly method. Therefore, it is important to develop reliable devices for point of care (PoC) virus detection [1]. The most common devices used for PoC virus screening are rapid antigen tests, which, however, show poor performance, as they are based on visual observation of the results, meaning they only provide qualitative results that cannot be automatically processed [2]. This gap is proposed to be filled by biosensors that can provide an electrical measurement, meaning a faster response time, improved sensitivity and the possibility of electronic processing of the results [3].

Even though the development of various biosensors has been reported [4–9], few have been used as complete SARS-CoV-2 screening devices.

In our previous work, we developed a label-free SARS-CoV-2 electrochemical biosensor based on the binding of the virus structural spike (S) protein to ACE2 protein [10]. ACE2 is immobilized in an interdigitated electrode (IDE) transducer [11], and the binding of the S protein (or the virus through S protein) to ACE2 [12] results in a change in the IDE electrical properties [13], hence its effective capacitance.

This paper presents the process of the evolution of the biosensor into a portable screening device for SARS-CoV-2 with smartphone readout. A portable microcontroller-based electronic readout circuit is developed, which performs effective capacitance measurements. The screening test results are acquired within 2 min and are made available via wireless transmission to a mobile application.

2. Materials and Methods

2.1. Biosensor Preparation

The biosensor preparation procedure has been previously reported [10]. In brief, gold interdigitated electrodes with an electrode length of 7 mm and an electrode surface area of 8.45 mm^2 were purchased from DropSens (Asturias, Spain). On top of the electrodes, ACE2 protein was immobilized. To verify the functionality of the device, S protein was placed on top of the biosensor, resulting in its binding to ACE2 and therefore a change in the electrical characteristics. ACE2 and S protein were purchased from InvivoGen (San Diego, CA, USA). All the chemicals used were purchased from Sigma-Aldrich (St. Louis, MO, USA).

2.2. Readout Circuit

An LCR meter was designed for the reading of the biosensor. The circuit could measure capacitance ranging from 1 pF up to 3 µF. It was based on the principle illustrated in Figure 1.

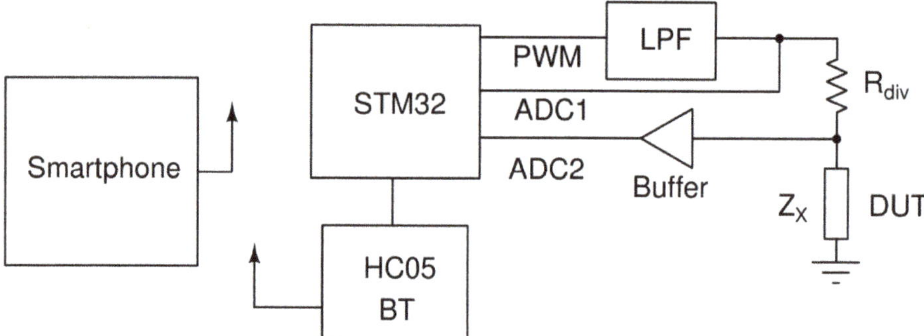

Figure 1. Working principle of the circuit.

A microcontroller unit (MCU) of the STM32 (STM32F103C8T6) family produced a high-frequency PWM signal that is then passed to a low pass filter (LPF). The LPF was a second-order Butterworth filter [14] with a cutoff frequency of 13 kHz. The output of the LPF, which was either 1 kHz or 10 kHz sinewave, drove a voltage divider consisting of a known resistor and the device under test (DUT). By measuring the amplitudes of the ADC1 and ADC2 voltages, as well as their phase difference, we could compute the impedance of DUT based on the following formulas:

$$Re(Z_x) = \frac{R_{div}|V_2|(V_1\,cos(\varphi) - |V_2|)}{V_1{}^2 - 2V_1\,|V_2|\,cos(\varphi) + |V_2|^2} \tag{1}$$

$$Im(Z_x) = \frac{V_1 R_{div}|V_2|\,sin(\varphi)}{V_1{}^2 - 2V_1\,|V_2|\,cos(\varphi) + |V_2|^2} \tag{2}$$

where V_1 is the voltage measured by ADC1, V_2 is the voltage measured by ADC2, φ is the phase of V_2 and the phase of V_1 is 0.

In order to reduce the noise of the measurement, the amplitudes and phases of the fundamental frequency were computed using the formula of Fourier transform. The result was then calculated by averaging the readings over 512 measurements and normalized by dividing every measurement with the maximum measured value.

2.3. Mobile Application

The deployment of a mobile application for easy, simple and direct access to the data provided by the sensor is considered necessary. In this project, a Bluetooth-based application for Android smartphones was designed. The Android platform includes support for the Bluetooth network stack, which allows a device to wirelessly exchange data with other Bluetooth-compatible devices over short distances. A Bluetooth transceiver module (HC-05), which is able to transmit data to the mobile application using the standard Bluetooth protocol, was added to the readout circuit. As a result, the Android application displayed the detection of S protein in the tested sample in real time. A mockup of the developed application is illustrated in Figure 2.

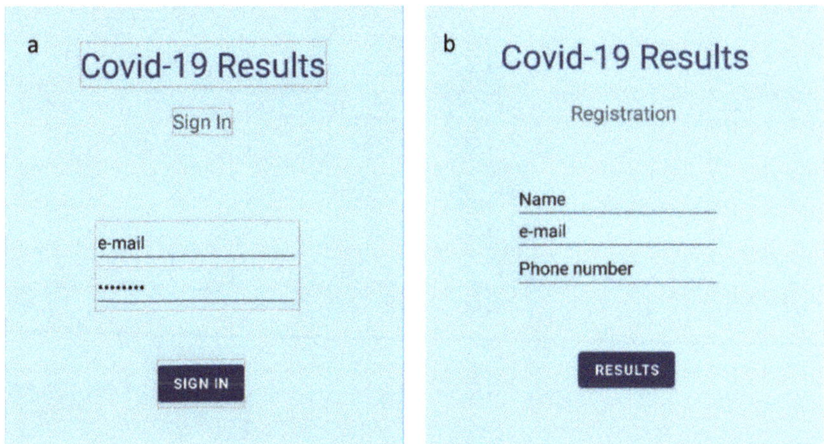

Figure 2. Mobile application's homepage; (**a**) sign in page; (**b**) registration page.

Google Android Studio was used for the development of the mobile application. Java was chosen to be used as the programming language, as it is the most popular language at the time of writing [15]. Regarding the graphical interface, the XML language was used.

First, the Android application allowed us to choose the appropriate Bluetooth device (i.e., the HC-05 module). When the connection was established, the application received the appropriate data packets in JSON format, sent from the readout circuit. Those data packets contained the outcome of the measurement, as well as the measured value and a timestamp. Finally, the result (positive or negative) was displayed on the screen.

3. Results and Discussion

3.1. Readout Circuit Calibration

The developed prototype board is illustrated in Figure 3. In the middle of the PCB, the *Blue Pill* STM32 development board was placed, along with the HC-05 BT module on the left, the LCD screen on top and the DIP switches for selecting the suitable range below it. The device under test was connected on the left and right female pins of the three-pin connector at the bottom. Lastly, on the bottom right, there are three buttons responsible for specifying frequency and current range and for performing open circuit calibration.

The measuring circuit was calibrated by measuring commercially available capacitors and resistors and comparing the results with those of specialized instruments. For capacitance measurement calibration, an Extech LCR Meter (Extech, model 380193) was used. For resistance measurements, a Keithley multimeter (Keithley 2000 Series) was used. Capacitance measurement results are illustrated in Table 1, and resistance measurement results are illustrated in Table 2. It is shown that the developed circuit can measure capacitance and resistance with high accuracy.

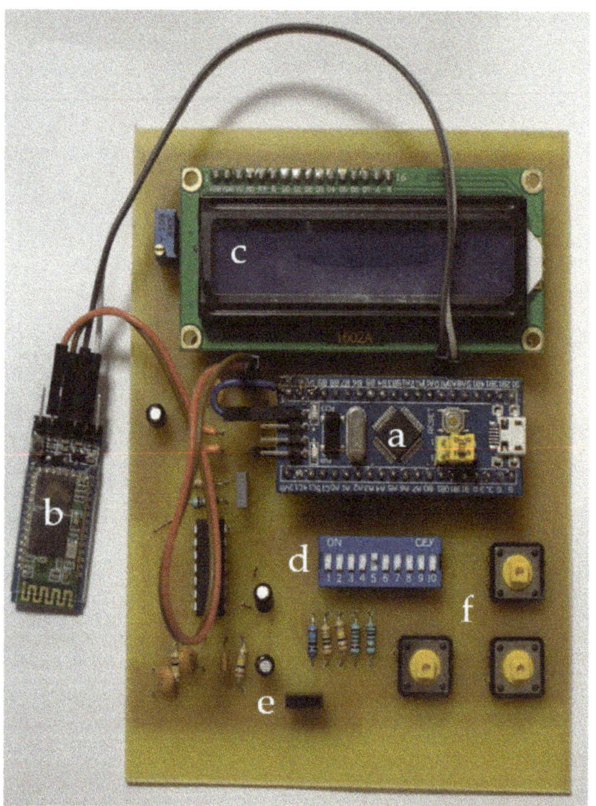

Figure 3. The prototype PCB; (**a**) Blue Pill STM32 development board; (**b**) HC-05 BT module; (**c**) LCD screen; (**d**) DIP switches; (**e**) input pins; (**f**) settings buttons.

Table 1. Capacitance measurements.

Nominal Value (F)	Extech 380193 (F)	Prototype PCB (F)	Relative Difference
10×10^{-12}	7.9×10^{-12}	10.5×10^{-12}	24.76%
100×10^{-12}	102.5×10^{-12}	104.5×10^{-12}	1.91%
1×10^{-9}	0.981×10^{-9}	0.999×10^{-9}	1.8%
2.2×10^{-9}	2.234×10^{-9}	2.245×10^{-9}	0.49%
10×10^{-9}	9.879×10^{-9}	9.918×10^{-9}	0.39%
100×10^{-9}	99.83×10^{-9}	103.6×10^{-9}	3.64%
1×10^{-6}	0.933×10^{-6}	0.935×10^{-6}	0.21%
2.2×10^{-6}	2.042×10^{-6}	2.038×10^{-6}	0.20%
3.3×10^{-6}	3.116×10^{-6}	3.12×10^{-6}	0.13%

Table 2. Resistance measurements.

Nominal Value (Ω)	Keithley 2000 (Ω)	Prototype PCB (Ω)	Relative Difference
100	99.5	99.93	0.43%
1×10^3	1×10^3	998.5	0.14%
4.7×10^3	4.6063×10^3	4.615×10^3	0.19%
10×10^3	10.008×10^3	10.023×10^3	0.15%
43×10^3	43.22×10^3	43.24×10^3	0.05%
100×10^3	99.21×10^3	99.3×10^3	0.09%
1×10^6	986.0×10^3	986.2×10^3	0.02%
6.8×10^6	6.843×10^6	6.841×10^6	0.03%
10×10^6	10.16×10^6	10.186×10^6	0.26%

3.2. Device Operation with S Protein

Experiments with S protein were conducted. A 20 µL solution containing S protein (2.5 ng/µL) in Phosphate-Buffered Saline (PBS) was placed on top of the sensor, and the effective capacitance change over time was monitored. A second experiment was conducted, where a 20 µL solution containing only PBS was placed on the sensor in order to calculate the blank solution response. The normalized capacitance change over time is illustrated in Figure 4.

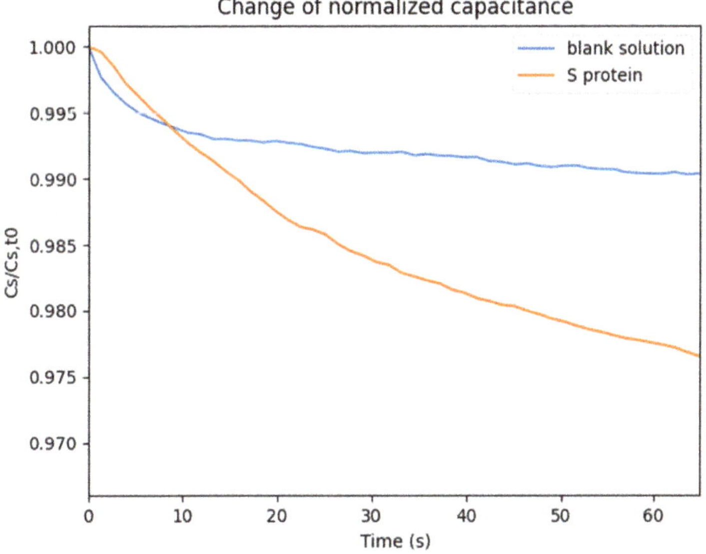

Figure 4. Normalized capacitance change over time.

The measurement procedure was the following: The user has to open the Android application and register by entering some personal information (Figure 2b) or sign in, if the user has previously registered (Figure 2a). Then, the testing procedure begins. After 60 s of measuring with a rate of 1 sample per second, the resulting value is transmitted to the Android application via Bluetooth. If the total capacitance change exceeds 1%, the test is listed as positive for the SARS-CoV-2 S protein and the user receives the appropriate response (Figure 5a). Otherwise, if the total capacitance change is below 1%, the test is listed as negative, and the user is informed as well (Figure 5b).

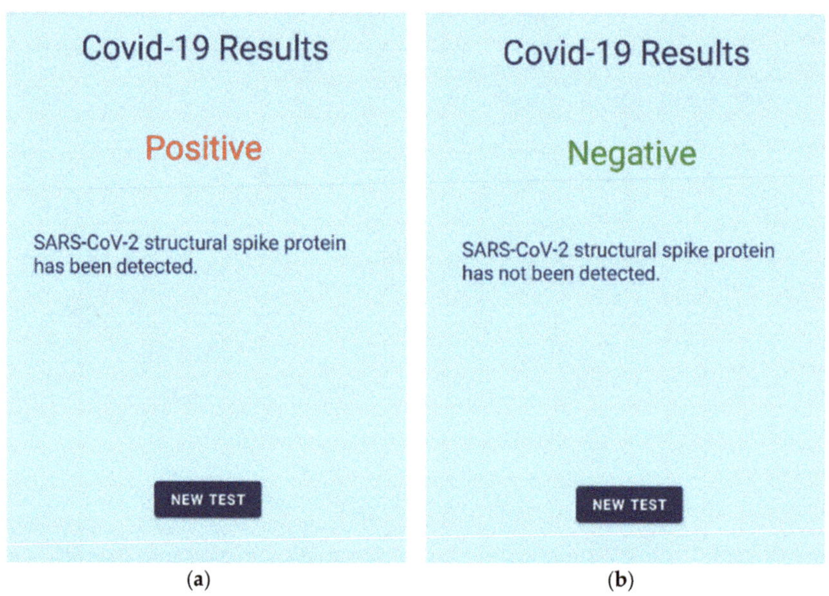

(**a**)　(**b**)

Figure 5. Mobile application results; (**a**) test positive to S protein; (**b**) test negative to S protein.

4. Conclusions

In this project, a SARS-CoV-2 S protein detecting device was developed using the ACE2-based capacitance sensor for rapid native SARS-CoV-2 detection [10]. The device consists of a microcontroller-based electronic circuit that, as shown, can measure capacitance and resistance change with high accuracy, and an Android application, where the test results are transmitted via Bluetooth. In our future work, we want to conduct more experiments and expand the device measurements to SARS-CoV-2 patient clinical samples that have been correlated with real-time PCR.

Author Contributions: A.G. conceptualized this study, performed experiments and wrote the first draft. K.A. implemented the electronic readout circuit and performed experiments. V.P. implemented the mobile application. S.A. performed experiments and coordinated the group. A.F. co-conceived the sensor with E.H. and supervised the biosensor preparation. E.H. co-conceived the sensor with A.F. and supervised the whole work. All authors have read and agreed to the published version of the manuscript.

Funding: This research was funded by Galenica SA (Kifisia, Greece) and Katharsis Technologies Inc. (Vancouver, BC, Canada).

Acknowledgments: We would like to thank the doctors of Konstantopoulio General Hospital E. Lampas and S. Patsilinakos for their guidance in this project.

Conflicts of Interest: The authors declare no conflict of interest. The funders had no role in the design of the study; in the collection, analyses, or interpretation of data; in the writing of the manuscript, or in the decision to publish the results.

References

1. Morales-Narváez, E.; Dincer, C. The impact of biosensing in a pandemic outbreak: COVID-19. *Biosens. Bioelectron.* **2020**, *163*, 112274. [CrossRef] [PubMed]
2. Eshghifar, N.; Busheri, A.; Shrestha, R.; Beqaj, S. Evaluation of Analytical Performance of Seven Rapid Antigen Detection Kits for Detection of SARS-CoV-2 Virus. *Int. J. Gen. Med.* **2021**, *14*, 435–440. [CrossRef] [PubMed]
3. Imran, S.; Ahmadi, S.; Kerman, K. Electrochemical biosensors for the detection of SARS-CoV-2 and other viruses. *Micromachines* **2021**, *12*, 174. [CrossRef] [PubMed]

4.   Mavrikou, S.; Tsekouras, V.; Hatziagapiou, K.; Paradeisi, F.; Bakakos, P.; Michos, A.; Koutsoukou, A.; Konstantellou, E.; Lambrou, G.I.; Koniari, E.; et al. Clinical Application of the Novel Cell-Based Biosensor for the Ultra-Rapid Detection of the SARS-CoV-2 S1 Spike Protein Antigen: A Practical Approach. *Biosensors* **2021**, *11*, 224. [CrossRef] [PubMed]

5.   Fathi-Hafshejani, P.; Azam, N.; Wang, L.; Kuroda, M.A.; Hamilton, M.C.; Hasim, S.; Mahjouri-Samani, M. Two-Dimensional-Material-Based Field-Effect Transistor Biosensor for Detecting COVID-19 Virus (SARS-CoV-2). *ACS Nano* **2021**, *15*, 11461–11469. [CrossRef] [PubMed]

6.   Rashed, M.Z.; Kopechek, J.A.; Priddy, M.C.; Hamorsky, K.T.; Palmer, K.E.; Mittal, N.; Valdez, J.; Flynn, J.; Williams, S.J. Rapid detection of SARS-CoV-2 antibodies using electrochemical impedance-based detector. *Biosens. Bioelectron.* **2020**, *171*, 112709. [CrossRef] [PubMed]

7.   Seo, G.; Lee, G.; Kim, M.J.; Baek, S.H.; Choi, M.; Ku, K.B.; Lee, C.S.; Jun, S.; Park, D.; Kim, H.G.; et al. Rapid Detection of COVID-19 Causative Virus (SARS-CoV-2) in Human Nasopharyngeal Swab Specimens Using Field-Effect Transistor-Based Biosensor. *ACS Nano* **2020**, *14*, 5135–5142. [CrossRef] [PubMed]

8.   Garg, M.; Sharma, A.L.; Singh, S. Advancement in biosensors for inflammatory biomarkers of SARS-CoV-2 during 2019–2020. *Biosens. Bioelectron.* **2021**, *171*, 112703. [CrossRef] [PubMed]

9.   Sharma, P.K.; Kim, E.-S.; Mishra, S.; Ganbold, E.; Seong, R.-S.; Kaushik, A.K.; Kim, N.-Y. Ultrasensitive and Reusable Graphene Oxide-Modified Double-Interdigitated Capacitive (DIDC) Sensing Chip for Detecting SARS-CoV-2. *ACS Sens.* **2021**, *6*, 3468–3476. [CrossRef] [PubMed]

10. Georgas, A.; Lampas, E.; Houhoula, D.P.; Skoufias, A.; Patsilinakos, S.; Tsafaridis, I.; Patrinos, G.P.; Adamopoulos, N.; Ferraro, A.; Hristoforou, E. ACE2-based capacitance sensor for rapid native SARS-CoV-2 detection in biological fluids and its correlation with real-time PCR. *Biosens. Bioelectron.* **2022**, *202*, 114021. [CrossRef] [PubMed]

11. Mazlan, N.S.; Ramli, M.M.; Abdullah, M.M.A.B.; Halin, D.S.C.; Isa, S.S.M.; Talip, L.F.A.; Danial, N.S.; Murad, S.A.Z. Interdigitated electrodes as impedance and capacitance biosensors: A review. *AIP Conf. Proc.* **2017**, *1885*, 20276. [CrossRef]

12. Lan, J.; Ge, J.; Yu, J.; Shan, S.; Zhou, H.; Fan, S.; Zhang, Q.; Shi, X.; Wang, Q.; Zhang, L.; et al. Structure of the SARS-CoV-2 spike receptor-binding domain bound to the ACE2 receptor. *Nature* **2020**, *581*, 215–220. [CrossRef] [PubMed]

13. Shang, J.; Ye, G.; Shi, K.; Wan, Y.; Luo, C.; Aihara, H.; Geng, Q.; Auerbach, A.; Li, F. Structural basis of receptor recognition by SARS-CoV-2. *Nature* **2020**, *581*, 221–224. [CrossRef] [PubMed]

14. Karki, J. *Active Low-Pass Filter Design*; Texas Instruments Application Report; Texas Instruments: Dallas, TX, USA, 2000.

15. Cheon, Y. Multiplatform Application Development for Android and Java. In Proceedings of the 2019 IEEE 17th International Conference on Software Engineering Research, Management and Applications (SERA), Honolulu, HI, USA, 29–31 May 2019; pp. 1–5. [CrossRef]

Proceeding Paper

MAX30102 Photometric Biosensor Coupled to ESP32-Webserver Capabilities for Continuous Point of Care Oxygen Saturation and Heartrate Monitoring †

Uriel Abe Contardi [1],*, Mateus Morikawa [1], Bruno Brunelli [1] and Douglas Vieira Thomaz [2]

1 Department of Engineering, Federal University of Technology of Parana, Parana 80230-901, Brazil; mateusmorikawa@select.eng.br (M.M.); brunobrunelli@select.eng.br (B.B.)
2 Department of Science, Biological and Environmental Technology, Università di Salento, 73100 Lecce, Italy; douglasvthomaz@gmail.com
* Correspondence: urielcontardi@alunos.utfpr.edu.br
† Presented at the 2nd International Electronic Conference on Biosensors, 14–18 February 2022; Available online: https://sciforum.net/event/IECB2022.

Abstract: Continuous monitoring is of upmost importance to manage emergency situations in healthcare. Therefore, we investigated the use of MAX30102, a commercial photometric biosensing module coupled to a ESP32 system-on-a-chip and its internet-of-things capabilities to continuously gather and process peripheral oxygen levels (SpO_2) and heartrates (HR) from users. Moreover, a user-friendly graphic interface was designed and implemented, and an anatomical case was 3D printed in thermoplastic polyester. Results showcased that the device functioned reliably, and according to literature describing photometric sensor functioning, thereby shedding light on the use of simple and affordable electronics for developing biosensing medical devices.

Keywords: internet-of-things; COVID-19; healthcare; medical device; bioelectronics

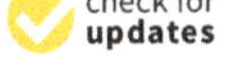

Citation: Contardi, U.A.; Morikawa, M.; Brunelli, B.; Thomaz, D.V. MAX30102 Photometric Biosensor Coupled to ESP32-Webserver Capabilities for Continuous Point of Care Oxygen Saturation and Heartrate Monitoring. *Eng. Proc.* **2022**, *16*, 9. https://doi.org/10.3390/IECB2022-11114

Academic Editors: Giovanna Marrazza and Sara Tombelli

Published: 14 October 2021

Publisher's Note: MDPI stays neutral with regard to jurisdictional claims in published maps and institutional affiliations.

Copyright: © 2021 by the authors. Licensee MDPI, Basel, Switzerland. This article is an open access article distributed under the terms and conditions of the Creative Commons Attribution (CC BY) license (https://creativecommons.org/licenses/by/4.0/).

1. Introduction

The development of innovative and affordable biosensing platforms for continuous biomarker monitoring is of the upmost importance for patient management during a healthcare crisis [1]. In this regard, several authors correlated the uninterrupted checking of patient signals to the improvement of clinical outcomes [2,3]. For instance, it has been reported that by following peripheral oxygen levels (SpO_2) and heartrates (HR), medical staff can prevent the aggravation of the symptoms of respiratory diseases, such as the one caused by the severe acute respiratory syndrome coronavirus 2 (SARS-CoV-2), which reaches a critical stage upon silent hypoxia [4].

Concerning the levels of SpO_2 and HR in diseases, it has been proved that respiratory impairment reduces oxygen saturation [4,5], whilst feedback mechanisms enhance the cardiac frequency by positive chronotropic effect [6,7]. This condition is known in many infectious respiratory diseases, such as SARS-CoV-2, but not limited to them. It is widely reported in literature that the drop in SpO_2 and ventricular tachycardia is a strong indication of chronic obstructive lung disease as well as other cardiopulmonary and circulatory ailments [8,9].

The monitoring of SpO_2 and HR in the clinical setting is performed by means of pulse oximeters, which evaluate the saturation through photometric means [10]. In this sense, the probe of the device houses two light-emitting diodes (LEDs), which emit light at 660 and 940 nm, as well as a photodiode [11]. This system is positioned so that the user's finger creates an interface between the LEDs and the photodiode, so that the light intensity captured by the photodiode changes according to the concentration of oxygen in the blood and due to the passage of blood through the finger. The resultant photoplethysmogram allows both the mensuration of SpO_2 and HR [12].

Although simple pulse oximeters are somewhat inexpensive, the need for health professionals to frequently check the outputs on the digital displays of these devices implies the requirement of constant surveillance. This can become a nuisance if human resources are limited, such as when there are many simultaneous hospitalizations, as in the current pandemic scenario [13]. In this sense, some works have interfaced pulse oximeters with wireless technologies, thereby allowing remote patient surveillance by health professionals [14–16].

The combination of internet-of-things (IoT)-based communication with artificial intelligence and classification tools in pulse oximeters has allowed for medical staff to better analyze patient status, evolution, and prediction of their clinical outcomes [17,18]. Nevertheless, even though promising, the products which employ wireless technologies are still costly, thereby hindering their acquisition by hospitals. In fact, this is further aggravated in developing nations due to taxation and currency exchange rates for imported electronic material [19], which therefore highlights the importance of developing innovative medical devices capable of affordably combining IoT communications and continuous monitoring.

Therefore, owing to the relevance in developing low-cost platforms to aid the remote surveillance of patient biomarkers, this work employed a commercial photometric module (i.e., MAX30102) and an inexpensive low-power system-on-a-chip (SoC) microcontroller (i.e., ESP32) in order to develop an open-open source IoT-based pulse oximeter to remotely monitor SpO_2 and HR continuously.

2. Methods

2.1. Materials

MAX30102 was used as the sensing module. This component is an integrated SpO_2 and HR monitor module for low-noise electronics with built-in ambient light rejection. MAX30102 functioning was fully validated and comprised of an optical module of 5.6 mm × 3.3 mm × 1.55 mm 14-pin with low-power HR monitor (<1 mW) and an ultra-low shutdown current of 0.7 µA, as well as robust motion artifact resilience and $-40\ ^\circ$C to $+85\ ^\circ$C operating temperature range. It could be supplied with a single 1.8 V source, or a separate 3.3 V [20,21]. Moreover, ESP32 was also used. This component was a 32-bit, low-cost, low-power SoC, which operates at 160 or 240 MHz, and had integrated IoT capabilities (Wi-Fi and dual-mode Bluetooth) [22]. The board used for the project were the LILYGO® TTGO T-Display that included the ESP32 and a 1.14-inch liquid crystal display (LCD); this board was also capable of running on a rechargeable lithium-ion battery. Furthermore, thermoplastic polyester (polylactic acid) (PLA) filaments were used in a custom 3D printer in order to build and assemble the case.

2.2. Circuit Design, Firmware, and Device Construction

The overall operation of the device consisted of the signal acquisition from the user's fingers by MAX30102. These signals were then transmitted to the ESP32 by SPI protocol. Moreover the MAX30102 was powered by the ESP32. The firmware in ESP32 SoC then allowed the information to be shown on the liquid crystal display (LCD), as well as to use ESP32 IoT capabilities to transmit data to the cloud (webserver). The firmware of the sensor platform herein described was developed using Arduino Integrated Development Environment (Arduino IDE). Both native and external libraries were used, such as Wi-Fi, SPI, Wire, as well as SparkFun electronics sensor MAX30102 library. The assembly of the device consisted of integrating both the sensing module and ESP32 SoC platform, and connecting MAX30102 and ESP32 with jumper wires made of copper and soldering iron. The jumper connections were 0.9 mm diameter. Furthermore, a case was designed to house each element of the device, and prototyped by 3D printing using yellow PLA filaments. The overall functioning of the device, and tridimensional rendering used for 3D printing are showcased in Figure 1.

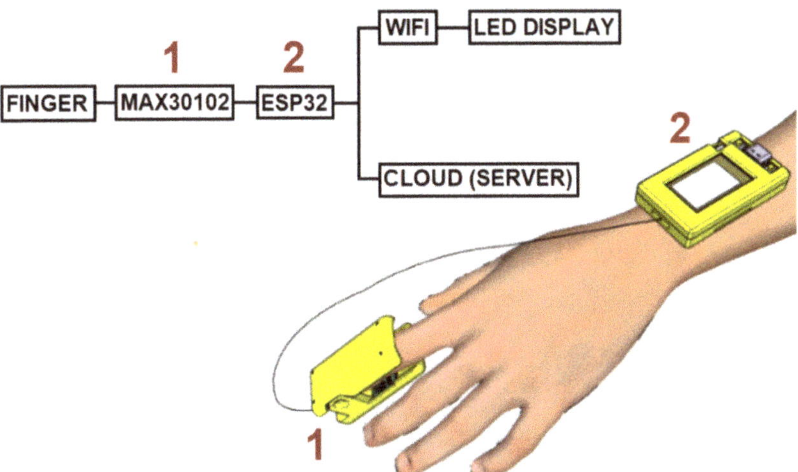

Figure 1. Outline of the device functioning, wherein the signals from patient's finger are gathered by MAX30102 and transmitted to ESP32, which both displays the information on an LCD display, as well as sends it to the cloud using the SoC IoT capabilities. (Rendering of the model used for the 3D printing of the device. ESP32 is housed in the wrist, while the MAX30102 is in the finger of the user. The connections between MAX30102 and ESP32 were performed with 0.9 mm diameter jumper wires.

3. Results and Discussion

After communicating MAX30102 and ESP32, the hardware was placed in the 3D-printed PLA case. Next, a universal serial bus (USB)-C cable was used to transfer the firmware to the device using Arduino IDE. The prototype is showcased in Figure 2.

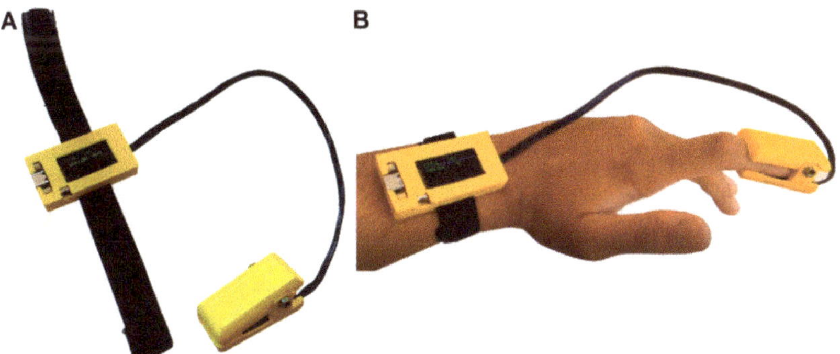

Figure 2. (**A**) Prototyped device. (**B**) Prototyped device on user's wrist and finger. ESP32 is located on the wrist portion of the device, whilst MAX30102 is located on the fingertip.

As showcased in Figure 2, the prototyped device presented the expected dimensions, being able to properly house ESP32 SoC and MAX30102 sensing module. Each part was firmly attached to the case; therefore, the user could freely move his hands without risking dislodging the components. Moreover, the clip at the finger portion worked well in attaching the fingertip of the user so that the sensor could touch the skin surface.

Furthermore, the readings of pulse oximetry were collected and compared to a standard pulse oximeter. Data was monitored both from the LCD, as well as ESP32 webserver, being the graphical product of the webserver results depicted in Figure 3.

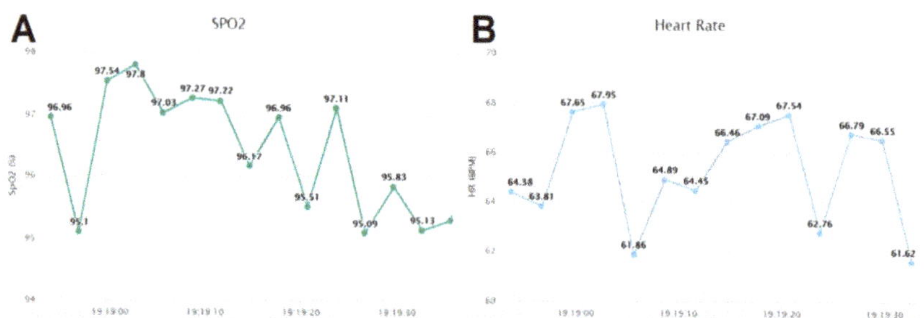

Figure 3. (**A**) SpO$_2$ and (**B**) HR Readings collected from ESP webserver. The values were in accordance to those of a standard pulse oximeter and were plotted in real time.

The readings could be performed both in the LCD at the wrist of the user, as well as remotely through a personal computer connected with ESP32 webserver. Both readings were the same and showcased values akin to those provided by a standard pulse oximeter. Considering that MAX30102 is a highly functional and reliable sensing module whose applicability in medical devices is widely reported and of acknowledged validation, the adequate functioning was an expected finding [13]. Moreover, the operation of the device followed the reliability described by other authors who communicated development platforms, such as the one herein used with MAX30102 [20,23]. In addition, the webserver capabilities of ESP32 were adequate to a single-user setting, taking into account the memory limits of this SoC.

Indeed, several developers described the easy integration of MAX30102 with ATmega328P-based Arduino and Tensilica Xtensa LX6 microprocessor-based ESPs [14,24]. Considering that the ESP32, such as the one herein used, allows ready IoT integration due to native wireless modules, its use is, therefore, more appropriate considering easiness of development and use. Moreover, ESP32 is considerably less bulky than Arduino development board, which allows for easier portability, such as by integrating the SoC on the wrist of the user for pulse oximetry purposes.

Nonetheless, the integration of ESP32 and MAX30102 has already been described by several developers and hobbyists due to its very easy reproduction. However, the integration of the IoT capabilities of ESP32 has not been often reported for biosensing purposes. In this regard, many reports described do-it-yourself prototype devices, which were of bench-top nature, as required direct communication with computers to operate and were not designed for portability. On the other hand, the device herein described is fully operational and IoT-integrated with simple programing, as well as portable and anatomic, due to the prototyped PLA case. Therefore, this work evidences how the use of simple and affordable electronics can assist the inexpensive develop IoT-based medical devices.

4. Conclusions

This work reported the use of MAX30102, a commercial photometric biosensing module coupled to ESP32 SoC and its IoT capabilities to continuously gather and process SpO$_2$ and HR from users. Results showcased that the device functioned reliably and according to literature describing photometric sensor functioning, thereby shedding light on the use of simple and affordable electronics for developing biosensing medical devices.

Author Contributions: U.A.C. conceptualized this study, performed experiments and wrote the first draft. Moreover, M.M. and B.B. contributed with experiments and writing the first draft of the manuscript. D.V.T. conceptualized the study, coordinated the group and wrote the final draft of the manuscript. All authors have read and agreed to the published version of the manuscript.

Funding: This research received no external funding.

Institutional Review Board Statement: Not applicable.

Informed Consent Statement: Not applicable.

Data Availability Statement: Not applicable.

Conflicts of Interest: The authors declare no conflict of interest.

References

1. Choi, J.R. Development of Point-of-Care Biosensors for COVID-19. *Front. Chem.* **2020**, *8*, 517. [CrossRef] [PubMed]
2. O'Carroll, O.; MacCann, R.; O'Reilly, A.; Dunican, E.M.; Feeney, E.R.; Ryan, S.; Cotter, A.; Mallon, P.W.; Keane, M.P.; Butler, M.W.; et al. Remote monitoring of oxygen saturation in individuals with COVID-19 pneumonia. *Eur. Respir. J.* **2020**, *56*, 2001492. [CrossRef] [PubMed]
3. Goodwin, R.; Aurora, T.; Gertz, J.; Gong, D.; Lykins, J.D. Remote oxygen monitoring for COVID-19 outpatient management. *Acad. Emerg. Med.* **2021**, *28*, 1.
4. Rahman, A.; Tabassum, T.; Araf, Y.; Al Nahid, A.; Ullah, M.A.; Hosen, M.J. Silent hypoxia in COVID-19: Pathomechanism and possible management strategy. *Mol. Biol. Rep.* **2021**, *48*, 3863–3869. [CrossRef]
5. Jiang, B.; Wei, H. Oxygen therapy strategies and techniques to treat hypoxia in COVID-19 patients. *Eur. Rev. Med. Pharmacol. Sci.* **2020**, *24*, 10239–10246.
6. Ståhlberg, M.; Reistam, U.; Fedorowski, A.; Villacorta, H.; Horiuchi, Y.; Bax, J.; Pitt, B.; Matskeplishvili, S.; Lüscher, T.F.; Weichert, I.; et al. Post-Covid-19 Tachycardia Syndrome: A distinct phenotype of Post-acute COVID-19 Syndrome. *Am. J. Med.* **2021**, *134*, 1451–1456. [CrossRef]
7. Raj, S.R.; Arnold, A.C.; Barboi, A.; Claydon, V.E.; Limberg, J.K.; Lucci, V.E.M.; Numan, M.; Peltier, A.; Snapper, H.; Vernino, S. Long-COVID postural tachycardia syndrome: An American Autonomic Society statement. *Clin. Auton. Res.* **2021**, *31*, 365–368. [CrossRef]
8. Long, B.; Brady, W.J.; Bridwell, R.E.; Ramzy, M.; Montrief, T.; Singh, M.; Gottlieb, M. Electrocardiographic manifestations of COVID-19. *Am. J. Emerg. Med.* **2021**, *41*, 96–103. [CrossRef]
9. Bandorski, D.; Höltgen, R.; Ghofrani, A.; Johnson, V.; Schmitt, J. Arrhythmias in patients with pulmonary hypertension and chronic lung disease. *Herzschrittmachertherapie Elektrophysiologie* **2019**, *30*, 234–239. [CrossRef]
10. Sangeeta, B.; Laxmi, S. A Real Time Analysis of PPG Signal for Measurement of SpO_2 and Pulse Rate. *Int. J. Comput. Appl.* **2011**, *36*, 45–50.
11. Nitzan, M.; Romem, A.; Koppel, R. Pulse oximetry: Fundamentals and technology update. *Med. Devices Evid. Res.* **2014**, *7*, 231. [CrossRef]
12. Elgendi, M.; Fletcher, R.; Liang, Y.; Howard, N.; Lovell, N.H.; Abbott, D.; Lim, K.; Ward, R. The use of photoplethysmography for assessing hypertension. *NPJ Digit. Med.* **2019**, *2*, 60. [CrossRef]
13. Suhartina, R.; Abuzairi, T. Pulse Oximeter Monitoring Bracelet for COVID-19 Patient using Seeeduino. *J. Ilm. Tek. Elektro Komput. Dan Inform.* **2021**, *7*, 81–87. [CrossRef]
14. Deivasigamani, S.; Narmadha, G.; Ramasamy, M.; Prasad, H.; Nair, P. Design of smart pulse oximeter using ATMEGA 328 microcontroller. *Int. J. Emerg. Technol.* **2020**, *11*, 696–700.
15. Hema, L.K.; Priya, R.M.; Indumathi, R. Design and Development of IOT Based Pulse Oximeter. *Int. J. Pure Appl. Math.* **2018**, *119*, 1863–1867.
16. Khairunnisa, S.; Gede, I.D.; Wisana, H.; Priyambada, I.; Nugraha, C.; Elektromedik, J.T. Rancang Bangun Pulse Oximeter Berbasis Iot (Internet of Things). *E-J. Poltekes Kemenkes Surabaya* **2018**, *1*, 28–32.
17. Zamanifar, A.; Nazemi, E.; Vahidi-Asl, M. DMP-IOT: A distributed movement prediction scheme for IOT health-care applications. *Comput. Electr. Eng.* **2017**, *58*, 310–326. [CrossRef]
18. Aldahiri, A.; Alrashed, B.; Hussain, W. Trends in Using IoT with Machine Learning in Health Prediction System. *Forecasting* **2021**, *3*, 12. [CrossRef]
19. Thomaz, D.V.; Contardi, U.A.; Morikawa, M.; dos Santos, P.A. Development of an affordable, portable and reliable voltametric platform for general purpose electroanalysis. *Microchem. J.* **2021**, *170*, 106756. [CrossRef]
20. Andika, I.P.A.; Rahmawati, T.; Mak'ruf, M.R. Pulse Oximeter Portable. *J. Electron. Electromed. Eng. Med. Inform.* **2019**, *1*, 28–32. [CrossRef]
21. Bento, A.C. An Experimental Survey with NodeMCU12e+Shield with Tft Nextion and MAX30102 Sensor. In Proceedings of the 11th Annual IEEE Information Technology, Electronics and Mobile Communication Conference, IEMCON 2020, Vancouver, BC, Canada, 4–7 November 2020.
22. Espressif Systems ESP32 Series Datasheet. Espressif Systems. 2019. Available online: https://www.espressif.com/sites/default/files/documentation/esp32_datasheet_en.pdf (accessed on 5 April 2022).
23. Ahmed, M.F.; Hasan, M.K.; Shahjalal, M.; Alam, M.M.; Jang, Y.M. Design and implementation of an OCC-based real-time heart rate and pulse-oxygen saturation monitoring system. *IEEE Access* **2020**, *8*, 198740–198747. [CrossRef]
24. Ramchandar Rao, P.; Rajendra Prasad, C.; Chitti, S.; Merugu, S.; Tarun Kumar, J. COVID-19 Patient Health Management System Using IoT. *LNNS* **2021**, *201*, 635–646.

Proceeding Paper

An Optical Fiber Sensor System for Uranium Detection in Water [†]

Nunzio Cennamo [1], Maria Pesavento [2,*], Daniele Merli [2], Antonella Profumo [2], Luigi Zeni [1] and Giancarla Alberti [2]

[1] Department of Engineering, University of Campania Luigi Vanvitelli, Via Roma 29, 81031 Aversa, Italy; nunzio.cennamo@unicampania.it (N.C.); luigi.zeni@unicampania.it (L.Z.)

[2] Department of Chemistry, University of Pavia, Via Taramelli 12, 27100 Pavia, Italy; daniele.merli@unipv.it (D.M.); antonella.profumo@unipv.it (A.P.); giancarla.alberti@unipv.it (G.A.)

* Correspondence: maria.pesavento@unipv.it

[†] Presented at the 2nd International Electronic Conference on Biosensors, 14–18 February 2022; Available online: https://sciforum.net/event/IECB2022.

Abstract: A simple to realize optical fiber sensor specific for uranium(VI) detection in water is reported to demonstrate the sensing approach's capability to determine uranyl (UO_2^{2+}) in water solution in the ppb range. The proposed sensor was obtained by combining a specific receptor layer for uranium to a gold thin film at which the surface plasmon resonance (SPR) phenomenon takes place via optical fiber. In particular, an SPR D-shaped plastic optical fiber (POF) probe was used for signal transduction. The proposed optical-chemical sensing method is attractive because, in principle, it can be applied directly in the field, giving an analytical response in a fast and not overly expensive manner.

Keywords: chemical sensors; surface plasmon resonance (SPR); plastic optical fibers (POFs); uranium; optical fiber sensors

Citation: Cennamo, N.; Pesavento, M.; Merli, D.; Profumo, A.; Zeni, L.; Alberti, G. An Optical Fiber Sensor System for Uranium Detection in Water. *Eng. Proc.* **2022**, *16*, 10. https://doi.org/10.3390/IECB2022-12296

Academic Editors: Giovanna Marrazza and Sara Tombelli

Published: 17 February 2022

Publisher's Note: MDPI stays neutral with regard to jurisdictional claims in published maps and institutional affiliations.

Copyright: © 2022 by the authors. Licensee MDPI, Basel, Switzerland. This article is an open access article distributed under the terms and conditions of the Creative Commons Attribution (CC BY) license (https://creativecommons.org/licenses/by/4.0/).

1. Introduction

The concentration of uranium in non-contaminated environmental waters is very low, often lower than 1 ppb in freshwater and around 3 ppb in sea water [1], so it is not of environmental or health concern. However, it can be much higher at contaminated sites. In particular, the by-product of the uranium enrichment process, depleted uranium, has been applied as armor-piercing ammunition in several international military conflicts because of its high density, hardness, and pyrophoric properties [2]. The testing and use of such ammunitions have led to the release of depleted uranium into the environment at several locations worldwide [3]. In this situation, the use of marker-free sensing devices for monitoring the uranium content of natural and contaminated waters can be of particular interest. Even if several analytical methods are currently available for the determination of uranium at the very low concentrations present in environmental waters, the proposed sensing method is attractive because, in principle, it can be applied on-site directly, giving an analytical response in a faster and not overly expensive manner.

In the present investigation, an optical fiber sensor for uranyl detection in water, based on surface plasmon resonance (SPR) transduction, is proposed. More specifically, an SPR D-shaped plastic optical fiber (POF) [4] has been covered by a specific receptor monolayer. This optical platform was developed by our research group more than ten years ago, and presents advantages in terms of low cost and small dimensions with respect to the classical Kretschman configuration [4]. It has already been applied to the detection of many different molecular species [5]. In some cases, it has been applied to detect different metal ions, such as copper(II) [6] and iron(III) [7]. In [6,7], strong and selective ligands for these metal ions were fixed at the SPR interface as a monolayer.

In this work, for the uranium(VI) detection, 11-mercaptoundecylphosphonic acid (MUPA) was used as a receptor monolayer, which is commercially available and possesses a sulfide terminal group that can be easily fixed to the gold surface of the SPR POF probe. In a previous research, the same receptor has been demonstrated to be suitable for a uranyl sensor with electrochemical transduction [8]. As a preliminary investigation, experimental results have been reported for a uranyl sensor based on the SPR POF platform combined with MUPA.

2. Materials and Methods

2.1. Reagents

For this study, 11-mercaptoundecylphosphonic acid (MUPA) was purchased from Sigma-Aldrich, as was uranyl (UO_2^{2+}) standard (uranyl nitrate, standard for ICPOES, 1000 mg L^{-1}). The lower concentration standards were obtained daily by dilution with Milli-Q water. Other reagents were always of the purest grade available and were used as received.

2.2. Instruments and Experimental Setup

The same experimental setup already used in [4,6,7] was employed here. The white light source (Halogen lamp HL-2000-LL manufactured by Ocean Optics, Dunedin, FL, USA) presents an emission range from 360 nm to 1700 nm, whereas the spectrometer FLAME-S-VIS-NIR-ES (manufactured by Ocean Optics, Dunedin, FL, USA) has a detection range from 350 nm to 1023 nm. The transmission spectra were displayed by Spectra Suite software (Ocean Optics, Dunedin, FL, USA). The spectra were normalized by the MATLAB software (MathWorks, Natick, MA, USA) using, as reference for normalization, the spectrum acquired with air as a surrounding medium over the gold surface (D-POF-bare) or MUPA-derivatized gold surface (D-POF-MUPA). All measurements were performed at 25 °C.

2.3. SPR-POF Platforms Preparation

The SPR platform was realized by exploiting a plastic optical fiber (POF) with a core of poly-methylmethacrylate (PMMA) of 980 μm and a cladding of fluorinated polymer of 10 μm (a total diameter of 1 mm), embedded in a resin support and erased as previously described [7] to produce a D-shaped region via polishing. The multilayer interface was realized as described for similar sensors. First, a Microposit S1813 photoresist, with a refractive index higher than that of the POF core, was deposited on the exposed POF by spinning at 6000 rpm for 60 s and polymerizing. Then, a gold film 60 nm thick was sputtered on the photoresist layer by a Bal-Tec SCD 500 machine.

The platforms produced in this way are indicated as SPR-D-POF-bare. The procedure is reported schematically in Figure 1.

2.4. MUPA Deposition

The selected ligand for uranium(VI) (i.e., MUPA) was immobilized as a monolayer on the gold surface, taking advantage of the presence of the sulfide group terminal group in MUPA according to the procedure previously described for an electrochemical sensor of uranyl [8]. The gold film was contacted overnight with a solution containing 2.5 mM MUPA in water. The modified POF platform was then abundantly rinsed with ethanol and Milli-Q water before use. The sensors produced in this way are indicated as SPR-D-POF-MUPA.

2.5. Measurements

About 50 μL of the aqueous sample solution were dropped over the flat sensing region and incubated at room temperature for ten minutes. The spectrum was registered and normalized, and the minimum transmission wavelength (the resonance wavelength) was evaluated.

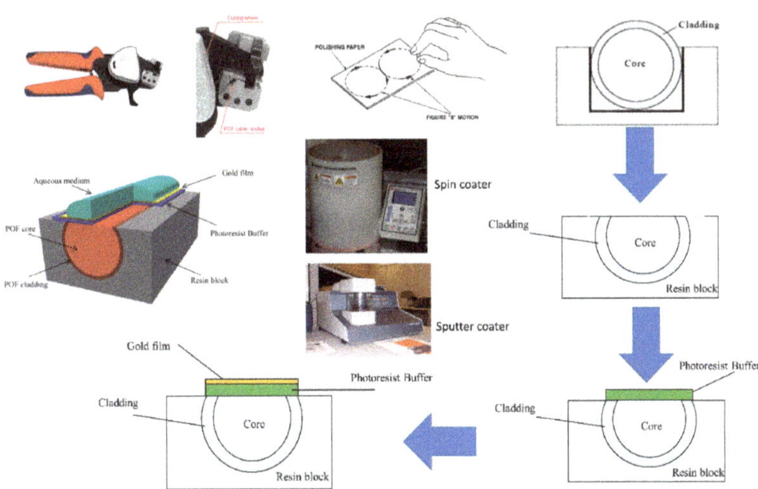

Figure 1. Preparation of the SPR platform based on D-shaped POF.

The quantity of analytical interest is the resonance wavelength variations with respect to the resonance wavelength of the blank solution (Δλris). The function Δλris versus uranyl concentration (c) were fitted by the Hill equation [9], using the software OriginPro (Origin Lab. Corp., Northampton, MA, USA). When the measured signal (Δλris (c)) is directly proportional to the concentration of the analyte adsorbed on the layer in contact with the gold surface, the response of the sensor is as follows:

$$\Delta\lambda\text{ris (c)} = \lambda c - \lambda 0 = \Delta\lambda\text{max} \cdot (c/(K + c)) \tag{1}$$

The symbols λc and λ0 indicate the resonance wavelength at 0 and c uranyl concentration, respectively. Δλmax is the value of the maximum resonance wavelength variation at increasing concentration of uranyl (c) (i.e., the value at saturation when c is much lower than K). Notice that according to the Langmuir adsorption model, K corresponds to the reciprocal of the affinity constant of the ion UO_2^{2+} for MUPA forming the thin receptor layer at the surface of gold. It can be evaluated, together with Δλmax, from Equation (1). Conversely, Equation (1) can be used as a standardization curve once the parameters are known.

3. Results

The optical sensitivity (bulk sensitivity) of the platforms was determined by measuring λris at the bare platform in dielectrics with different refractive indexes, as previously reported [4]. Water-glycerol solutions were used for this experiment, obtaining a sensitivity of 2500 nm RIU^{-1} in the wavelength range considered.

The presence of the receptor layer on the gold surface was demonstrated by the SPR method by comparing the transmission spectrum of the bare platform (SPR-D-POF-bare) in water with that of the MUPA-derivatized platform (SPR-D-POF-MUPA) in water. An example is reported in Figure 2, in which it is seen that the spectrum of SPR-D-POF-MUPA normalized on the spectrum in air is similar to that of the SPR-D-POF-bare normalized on the spectrum in air, but with the resonance wavelength shifted to higher values. This clearly indicates the presence of some receptor molecules at the gold surface producing a refractive index higher than that of pure water.

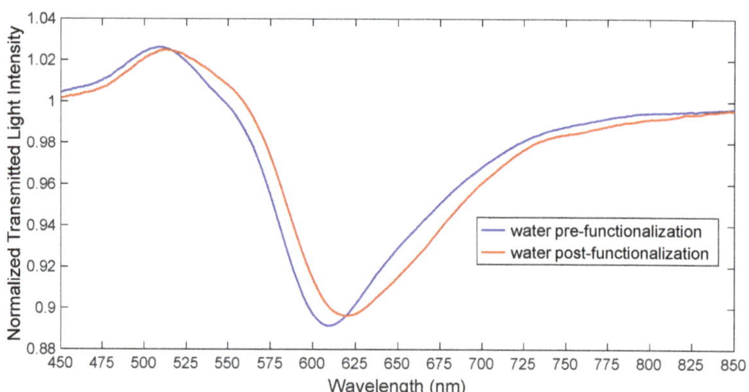

Figure 2. Transmission spectra of SPR-D-POF-bare and SPR-D-POF-MUPA in water normalized to the spectrum of the corresponding platform in air.

The spectra of SPR-D-POF-MUPA in 0.1 M NaNO$_3$ solution containing different concentrations of uranium(VI), normalized on the corresponding spectra in air, are reported in Figure 3. As shown in Figure 3, when the analyte concentration increases, the resonance wavelength increases and shifts to the right; in fact, the refractive index in contact with the gold nanofilm increases with the uranium concentration.

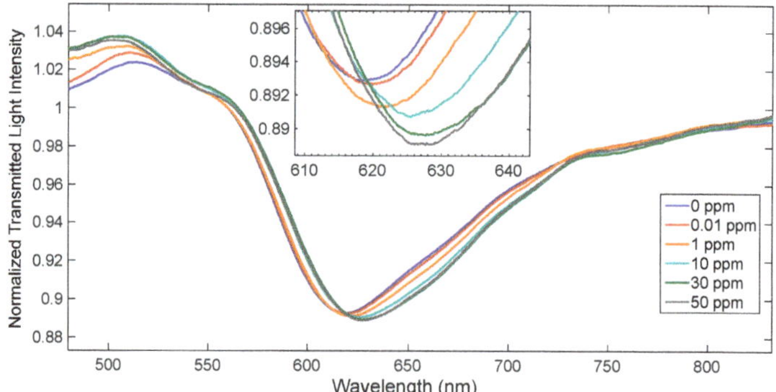

Figure 3. SPR spectra obtained at different uranium concentrations (in mg L^{-1} (ppm)).

4. Conclusions

The developed SPR sensor for uranium(VI) is based on an optical platform that is fast and easy to prepare, including the deposition of the receptor, a monolayer of a well-known commercially available complexing agent for uranyl. When deposited on the gold surface, its affinity for uranyl is sufficiently high to reach a LOD of a few µg L^{-1}, suitable for detection in slightly contaminated waters. However, this limit could be improved by optimizing the procedure of the monolayer deposition in order to have a higher density of complexing sites at the sensor surface or by selecting a stronger ligand.

Author Contributions: Conceptualization: M.P. and N.C.; methodology: M.P. and N.C.; validation: M.P., D.M., G.A., A.P., L.Z. and N.C.; formal analysis: M.P., D.M., G.A., A.P., L.Z. and N.C.; investigation: M.P., D.M., G.A., A.P., L.Z. and N.C.; writing—original draft preparation: M.P., D.M., G.A., A.P., L.Z. and N.C.; writing—review and editing: M.P., D.M., G.A., A.P., L.Z. and N.C.; supervision: M.P. All authors have read and agreed to the published version of the manuscript.

Funding: This research received no external funding.

Institutional Review Board Statement: Not applicable.

Informed Consent Statement: Not applicable.

Data Availability Statement: The data are available on reasonable request from the corresponding author.

Conflicts of Interest: The authors declare no conflict of interest.

References

1. Alberti, G.; Biesuz, R.; Pesavento, M. Determination of the total concentration and speciation of Uranium in natural waters by the Resin Titration method. *Microchem. J.* **2007**, *86*, 166–173. [CrossRef]
2. Danesi, P.R.; Bleise, A.; Burkart, W.; Cabianca, T. Isotopic composition and origin of uranium and plutonium in selected soil samples collected in Kosovo. *J. Environ. Radioact.* **2003**, *64*, 121–131. [CrossRef]
3. Saleh, I.H.; Abdel-Halim, A.A. Determination of depleted uranium using a high-resolution gamma-ray spectrometer and its applications in soil and sediments. *J. Taibah Univ. Sci.* **2016**, *10*, 205–211. [CrossRef]
4. Cennamo, N.; Massarotti, D.; Conte, L.; Zeni, L. Low Cost Sensors Based on SPR in a Plastic Optical Fiber for Biosensor Implementation. *Sensors* **2011**, *11*, 11752–11760. [CrossRef] [PubMed]
5. Cennamo, N.; Pesavento, M.; Zeni, L. A review on simple and highly sensitive plastic optical fiber probes for bio-chemical sensing. *Sens. Actuators B Chem.* **2021**, *331*, 129393. [CrossRef]
6. Pesavento, M.; Profumo, A.; Merli, D.; Cucca, L.; Zeni, L.; Cennamo, N. An Optical Fiber Chemical Sensor for the Detection of Copper(II) in Drinking Water. *Sensors* **2019**, *19*, 5246. [CrossRef] [PubMed]
7. Cennamo, N.; Alberti, G.; Pesavento, M.; D'Agostino, G.; Quattrini, F.; Biesuz, R.; Zeni, L. A Simple Small Size and Low Cost Sensor Based on Surface Plasmon Resonance for Selective Detection of Fe(III). *Sensors* **2014**, *14*, 4657–4661. [CrossRef] [PubMed]
8. Merli, D.; Protti, S.; Labò, M.; Pesavento, M.; Profumo, A. A ω-mercaptoundecylphosphonic acid chemically modied gold electrode for uranium determination in waters in presence of organic matter. *Talanta* **2016**, *151*, 119–125. [CrossRef] [PubMed]
9. Garcia-Calzon, J.A.; Diaz-Garcia, M.E. Characterization of binding sites in molecularly imprinted polymers. *Sens. Actuators B Chem.* **2007**, *123*, 1180–1194. [CrossRef]

engineering proceedings

MDPI

Proceeding Paper

Trends and Recent Patents on Cellulose-Based Biosensors [†]

Ahmed Fatimi [1,2]

1 ERSIC, Polydisciplinary Faculty, Sultan Moulay Slimane University (USMS), P.O. Box 592, Mghila, Beni-Mellal 23000, Morocco; a.fatimi@usms.ma
2 Department of Chemistry, Polydisciplinary Faculty, Sultan Moulay Slimane University (USMS), P.O. Box 592, Mghila, Beni-Mellal 23000, Morocco
† Presented at the 2nd International Electronic Conference on Biosensors, 14–18 February 2022; Available online: https://sciforum.net/event/IECB2022.

Abstract: Research on cellulose and its derivatives as biosensors is developing rapidly through the innovation and improvement of materials, chemical synthesis, and methods of preparation and formulation. This study presents the state of the art by introducing what has been innovated and patented concerning cellulose-based biosensors between 2010 and 2020. More specifically, this form of patent analysis encapsulates information that could be used as a reference by researchers in the fields of biosensors and cellulose-based biosensing platforms, as well as those interested especially in cellulose and its derivatives. As a result of this study, a total of 241 patent documents related to cellulose-based biosensors were found. The United States leads the patent race in this sector. Based on patent classifications, most patents and inventions are intended for chemical analysis of biological materials and testing involving biospecific ligand binding methods, as well as measuring or testing apparatus with condition measuring or sensing means. Research and development are based on the investigating or analyzing of materials by the use of electric or electrochemical means, as well as nanotechnology for interacting, sensing, or actuating, which are concentrated in most patents.

Keywords: cellulose; biosensors; patents; innovation

check for **updates**

Citation: Fatimi, A. Trends and Recent Patents on Cellulose-Based Biosensors. *Eng. Proc.* **2022**, *16*, 12. https://doi.org/10.3390/IECB2022-12253

Academic Editors: Giovanna Marrazza and Sara Tombelli

Published: 14 February 2022

Publisher's Note: MDPI stays neutral with regard to jurisdictional claims in published maps and institutional affiliations.

Copyright: © 2022 by the author. Licensee MDPI, Basel, Switzerland. This article is an open access article distributed under the terms and conditions of the Creative Commons Attribution (CC BY) license (https://creativecommons.org/licenses/by/4.0/).

1. Introduction

Materials used for fabricating biomedical devices, such as implantable biosensors, need to possess appropriate physical, chemical, and biological properties, depending on specific circumstances [1]. Among a variety of materials commonly used as biosensing platforms, cellulose and its derivatives have gained considerable attention [2,3].

Cellulose and its derivatives exhibit biologically appropriate properties which make them suitable for biomedical applications [4–6]. Due to its promising physical and biological characteristics as well as chemical structure, cellulose has been shown to be a multifunctional material, providing a high-quality platform to complete the immobilization process of bioactive molecules in biosensors [7]. Nevertheless, some researchers have investigated methods to improve cellulose properties to meet biosensing requirements and then facilitate the applicability of cellulose in biosensing [2,8].

The first patent application concerning cellulose-based biosensors was filed in 1987 and then granted in 1991 [9]. Through this patent, Godfrey invented methods of producing suitable polymers for use as optical structures. More specifically, the inventor proposed new ways of applying an optical coating to the surface of a polymer-coated structure to make surfaces suitable for use in biosensors. A claimed method in this patent suggested that the polymer was cellulose nitrate [9].

Research on cellulose and its derivatives as biosensors is developing rapidly through the innovation and improvement of raw materials, chemical synthesis, and methods of preparation and formulation, with more than 200 organizations around the world currently involved in patent activity and filings concerning cellulose-based biosensors. This trend is supported by cellulose's numerous advantages in biosensing and biomedical applications. This is also evident from the increase in the number of patent applications filed each year worldwide in the field of cellulose-based biosensor research and development [10].

This study, in the form of patent analysis, presents the state of the art by introducing what has been patented in relation to cellulose-based biosensors between 2010 and 2020. Furthermore, detailed charts have been provided by determining publication year, classifications, inventors, applicants, owners, and jurisdictions. Finally, a selection of relevant granted patents and patent applications is proposed to demonstrate the innovation and improvement of these cellulose-based biosensors.

2. Methods

2.1. Resources and Research Methods

Two databases have been used in this study: the Patentscope search service of the World Intellectual Property Organization (WIPO) [10,11] and The Lens patent data set [12]. Different keywords related to cellulose-based biosensors were used, and patent documents were searched according to title, abstract, and claims. The results were then filtered to include only patent documents with a publication date between 1 January 2010 and 31 December 2020.

2.2. Analysis of the Patentability of Cellulose-Based Biosensors

As a result of the search, 241 patent documents were found. Generally, this encompassed patent applications and granted patents. In relation to cellulose-based biosensors, the found patent documents have been classified as 204 patent applications and 37 granted patents. The detailed list of these patent documents is presented in Table S1 (Supplementary Materials).

Hereinafter, the state of the art will be reviewed by introducing what has been patented concerning cellulose-based biosensors. A detailed analysis of the patentability will be provided, with emphasis on the used cellulose and its derivatives, testing involving biospecific ligand binding methods, and interacting, sensing, or actuating, following the publication year, patent classifications, inventors, applicants, owners, and jurisdictions.

3. Results and Discussion

3.1. Publication Year

The publication year is the year in which a patent document (i.e., patent application, granted patent, etc.) was published and made available to the public [13].

Concerning cellulose-based biosensors, our search found 241 patent documents between 2010 and 2020. The results encompassed 204 patent applications and 37 granted patents. Our search found five registered patent documents for the year 2010. In contrast, the year 2020 recorded 34 patent documents. The maximum number of granted patents (eight) was recorded in 2018. Furthermore, the year 2020 was the year with the maximum patent applications or patent documents, with 27 and 34, respectively, (Figure 1).

3.2. Patent Classifications

The International Patent Classification (IPC) is a code-based hierarchical system that separates all technological domains into sections, classes, subclasses, groups, and subgroups. It is a global classification system that provides standardized data for categorizing inventions and assessing their technological distinctiveness [14,15].

Concerning cellulose-based biosensors, the top 10 IPC codes between 2010 and 2020 are presented in Figure 2. The most-recorded IPC code corresponds to G01N27/327, which is a subgroup of investigating or analyzing materials by the use of electric or electrochemical means, such as biochemical electrodes. This subgroup alone recorded 60 patent documents. The second most-recorded IPC code corresponds to C12Q1/00, which is a group of measuring or testing processes involving enzymes, nucleic acids, or microorganisms. It concerns measuring or testing apparatus with condition-measuring or sensing means. This group has 45 recorded patent documents. For more details concerning this top 10, a description of each IPC code is shown in Table 1.

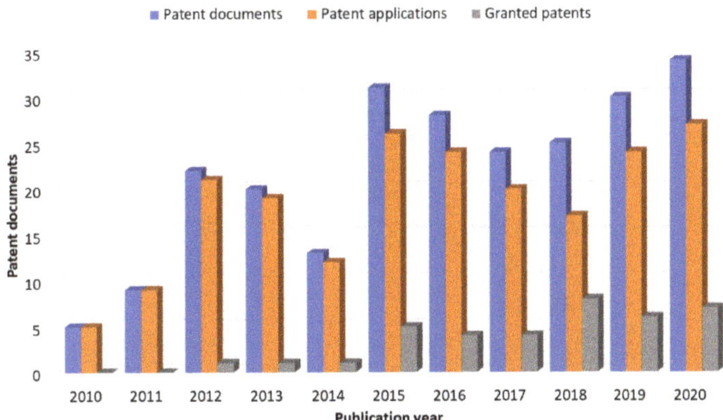

Figure 1. Evolution of patent documents (i.e., patent applications and granted patents) as a function of the published date of cellulose-based biosensors between 2010 and 2020.

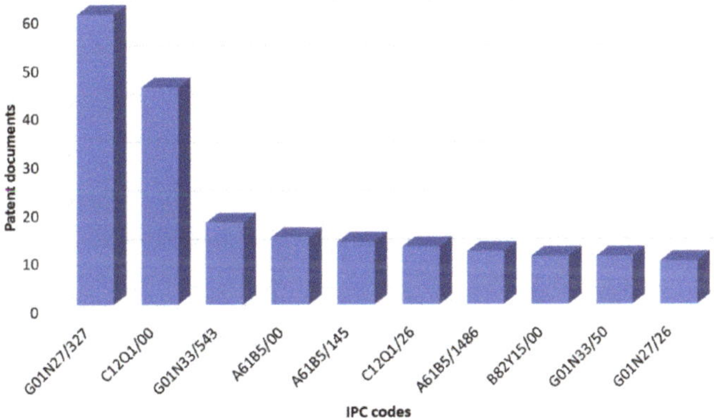

Figure 2. IPC codes (top 10) of the resulted patents as a function of patent documents of cellulose-based biosensors between 2010 and 2020.

3.3. Inventors

An inventor is a natural person designated for a patent application [13].

Concerning cellulose-based biosensors, the top 10 inventors between 2010 and 2020 are presented in Figure 3. Jung Sung-Kwon from the Republic of Korea is ranked as the first inventor who has recorded 15 patent documents. The inventors Cha Geun Sig and Nam Hakhyun, from the Republic of Korea, tied for second place with 11 patent documents each.

Table 1. Meaning of IPC codes concerning the resultant patents of cellulose-based biosensors [14].

IPC Codes	Description
G01N27/327	Investigating or analyzing materials by the use of electric or electrochemical means, such as biochemical electrodes.
C12Q1/00	Measuring or testing processes involving enzymes, nucleic acids or microorganisms (measuring or testing apparatus with condition-measuring or sensing means).
G01N33/543	Immunoassay and biospecific binding assay with an insoluble carrier for immobilizing immunochemicals.
A61B5/00	Measuring for diagnostic purposes.
A61B5/145	Measuring characteristics of blood in vivo (e.g., gas concentration, pH-value, etc.).
C12Q1/26	Measuring or testing processes involving oxidoreductase.
A61B5/1486	Measuring characteristics of blood in vivo by using enzyme electrodes (e.g., with immobilized oxidase).
B82Y15/00	Nanotechnology for interacting, sensing or actuating.
G01N33/50	Chemical analysis of biological material (e.g., blood, urine, etc.); Testing involving biospecific ligand binding methods; Immunological testing.
G01N27/26	Investigating or analyzing materials by the use of electric or electrochemical means by investigating electrochemical variables or by using electrolysis or electrophoresis.

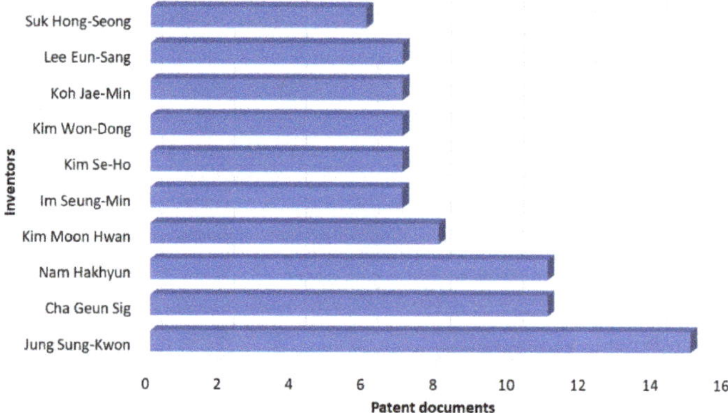

Figure 3. Inventors (top 10) of the resulted patents as a function of patent documents of cellulose-based biosensors between 2010 and 2020.

All found patent documents by the above three inventors concern the healthcare companies i-Sens Inc., (Seoul, Republic of Korea) and Osang Healthcare Co. Ltd., (Dongan-gu Anyang, Republic of Korea) as applicants and/or owners (Figures 4 and 5). i-Sens Inc. is a healthcare company that develops, produces, and distributes medical devices. The company's main products include blood glucose monitoring systems, electrolyte and gas analyzers, and immunosensors [16]. However, Osang Healthcare Co. Ltd. is a healthcare company that manufactures and distributes medical supplies. The company produces diagnostic biosensors for blood glucose measurements, cholesterol-measuring devices, and other related products [17].

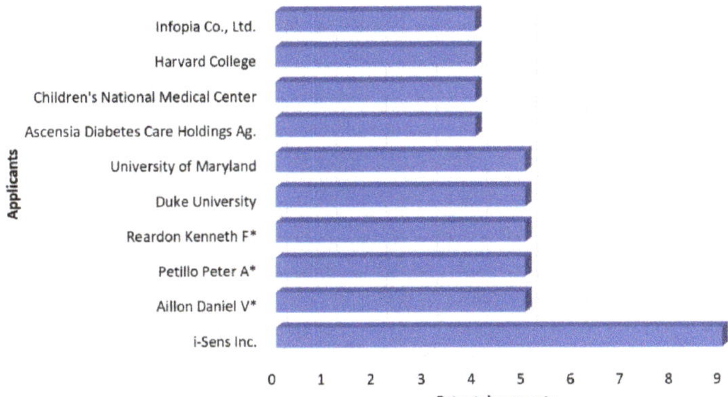

Figure 4. Applicants (top 10) of the resulted patents as a function of patent documents of cellulose-based biosensors between 2010 and 2020. * The applicant is a natural person.

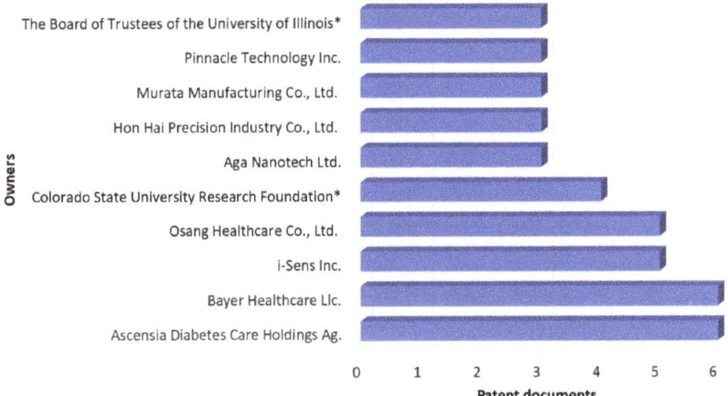

Figure 5. Owners (top 10) of the resulted patents as a function of patent documents of cellulose-based biosensors between 2010 and 2020. * The owner is a foundation or a governing body.

3.4. Applicants

In the case of a patent application, an applicant is a natural person or a legal entity that has filed the application [13].

Concerning cellulose-based biosensors, the top 10 applicants between 2010 and 2020 are presented in Figure 4. Regarding this top 10, all applicants are considered as people or organizations (companies and universities). As a legal entity, the healthcare company i-Sens Inc. (Seoul, Republic of Korea) is ranked as the first applicant that has recorded nine patent documents.

3.5. Owners

An owner is a natural person or a legal entity to whom the inventor or applicant has assigned the right to a patent [18,19].

Concerning cellulose-based biosensors, the top 10 owners between 2010 and 2020 are presented in Figure 5. The healthcare companies Ascensia Diabetes Care Holdings Ag. (Basel, Switzerland) and Bayer Healthcare Llc. (Emeryville, CA, USA), as legal entities, are ranked as the first owners who have recorded 6 patent documents each.

3.6. Jurisdictions

A patent application can be filed at the appropriate patent office under whose jurisdiction the applicant normally resides, has his domicile, or has a place of business, or at the place from where the invention actually originated. In several cases, related patent applications can be filed in multiple jurisdictions [20,21].

Concerning cellulose-based biosensors, the top 10 jurisdictions between 2010 and 2020 are presented in Figure 6. The United States, through the USPTO (United States Patent and Trademark Office), encompassed 110 patent documents with a higher patent contribution per total of ~46%; the global system for filing patent applications, known as the Patent Cooperation Treaty (PCT) and administered by WIPO, encompassed 73 patent documents with a patent contribution per total of ~30%; China, through the CNIPA (China National Intellectual Property Administration), encompassed 27 patent documents with a patent contribution per total of ~11%; finally, the EPO (European Patent Office), through which patent applications are filed regionally (Europe), encompassed 23 patent documents with a patent contribution per total of ~10%.

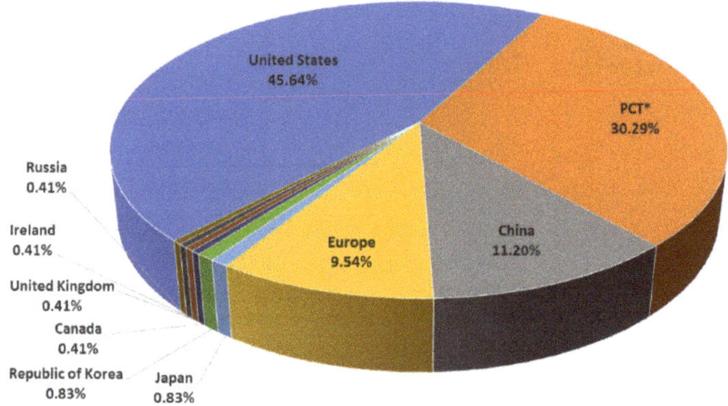

Figure 6. Patent contribution (%) as a function of jurisdiction of filed patent applications and granted patents of cellulose-based biosensors between 2010 and 2020. * The global system for filing patent applications, known as the Patent Cooperation Treaty (PCT) and administered by WIPO.

4. Selection of Relevant Patents

Cellulose-based biosensors are a technological innovation in the field of nanotechnology for sensing and analyzing by the use of electrochemical means. These cellulose-based biosensors are developing rapidly through the improvement of materials, chemical synthesis, and methods of preparation and formulation.

Table 2 presents examples of innovation and utilization of cellulose-based biosensors as demonstrated by inventions and patents. To clarify the real status of such patents in this area, the ones described herein are the most relevant patents focused on cellulose-based biosensors between 2010 and 2020. The choice of these selected patent documents was based on the most prolific countries patenting cellulose-based biosensors, as well as the patenting levels (national, regional, and international). However, the detailed list of all patent documents is presented in Table S1 (Supplementary Materials).

Table 2. Selection of relevant patents for cellulose-based biosensors between 2010 and 2020.

Patent	Publication	Title	References
WO 2010/124270 A1 Application No: 2010032329	2010-10-28	Functionalized polymer biosensor	[22]
KR 20120050359 A Application No: 20100111810	2012-05-18	Disposable biosensors made with tin oxide-cellulose nanocomposite, manufacturing method of the same and measuring method using the same	[23]
US 8,349,168 B2 Application No: 201213433779	2013-01-08	Determination of blood glucose in a small volume and in a short test time using a chemical coating including binders and very short read potentials	[24]
EP 2,821,497 A2 Application No: 14175390	2015-01-07	Reagent composition for biosensor and biosensor having the same	[25]
US 2018/0271423 A1 Application No: 201615764627	2018-09-27	A biosensor device to detect target analytes, in situ, in vivo, and/or in real time, and methods of making and using the same	[26]
CN 111,839,532 A Application No: 202010675959	2020-10-30	Flexible epidermal electrochemical biosensor based on conductive hydrogel	[27]

In 2010, Baker, et al., invented a novel sensor mechanism based on the aggregation of nanoparticles for target molecule detection and quantification. The signal-enhanced biosensor system, according to this patent, was made with a substrate that was selected from a group consisting of, among other things, a cellulose substrate or a nitrocellulose substrate [22].

In 2012, Kim invented a disposable biosensor made of tin oxide-cellulose nanocomposite. The invented biosensor was biodegradable, biocompatible, and flexible. It was obtained by coating tin oxide on the surface of a regenerated cellulose film and fixing the biochemical receptor (e.g., glucose oxidase, glutaminase, asparaginase, penicillinase, nitrate reductase, etc.) by a conventional method, such as encapsulation, covalent binding, crosslinking, and adsorption [23].

In 2013, Wilsey invented a method for determining the concentration of glucose in a blood sample by using sample volumes of less than a microliter, and test times within about eight seconds of the application of the sample. The invention is related to an electrochemical sensor comprising an array of micro-electrodes disposed on a flexible substrate. The sensor includes a chemical coating made up of a binder and contains a set of chemicals (e.g., methylcellulose, hydroxyethylcellulose, carboxymethylcellulose, microcrystalline cellulose, etc.) that react to produce an electroactive reaction product [24].

In 2015, Jeon, et al., invented a reagent composition for a biosensor having high sensitivity, which is capable of improving analysis linearity and detecting the concentration of a small amount of the analyte, such as glucose, by reacting with an oxidoreductase. The invented biosensor reagent composition, according to patent claims, encompasses, amongst other ingredients, a water-soluble polymer such as hydroxyethyl cellulose, hydroxypropyl cellulose, carboxy methyl cellulose, and cellulose acetate [25].

In 2018, Agarwal, et al., invented a biosensor for detecting the presence of a target analyte in a sample. Disclosed in the patent is a biosensor device for the real-time detection of a target analyte that includes a receptor component operatively connected to a transducer component which is adapted to interpret and transmit a detectable signal. The claimed biosensor uses a sensing element that comprises a polymer and was selected from a group consisting of, among others, carboxymethyl cellulose and derivatives [26].

In 2020, Pu et al., invented a flexible skin electrochemical biosensor based on a conductive, cellulose-based hydrogel. The newly invented biosensor comprised a reference electrode, a counter electrode, a working electrode, and a signal wire, which were arranged on the conductive hydrogel film substrate layer. The formulation of the conductive hydrogel film substrate layer comprised mixing a zwitterionic polymer monomer, a nano-cellulose solution, a thermal initiator, and a crosslinking agent. The working electrode was used for measuring biomolecules, and the signal wire was connected to external equipment. The invention improved the requirements for target molecule detection performance and wearing comfort [27].

5. Conclusions

This study provided patentability and patent analysis of cellulose-based biosensors. The detailed analysis concerned the period between 2010 and 2020. The publication year, 2020, was the year with the highest patent registration, with 34 patent documents. The United States was ranked first with 110 patent documents. Based on the IPC codes, all filed patents concerned chemical analysis of biological materials, and testing involving biospecific ligand binding methods, as well as measuring or testing apparatus with condition-measuring or sensing means. In addition, this study demonstrated that the inventions included in the patents concerned cellulose formulation and the process for preparing it, as well as devices and apparatus for manufacturing cellulose-based biosensors. The knowledge clusters and expert driving factors of this patent analysis indicate that the research and development were based on the investigating or analyzing of materials by the use of electric or electrochemical means, as well as nanotechnology for interacting, sensing, or actuating, which was concentrated in most patents.

Supplementary Materials: The following supporting information can be downloaded at: https://www.mdpi.com/article/10.3390/IECB2022-12253/s1, Table S1: Detailed list of all patent documents of cellulose-based biosensors between 2010 and 2020.

Funding: This research received no external funding.

Institutional Review Board Statement: Not applicable.

Informed Consent Statement: Not applicable.

Data Availability Statement: The data presented in this study are available within this article content.

Acknowledgments: The author acknowledges the World Intellectual Property Organization for the Patentscope search service and the Cambia Institute for The Lens patent data set used in this study.

Conflicts of Interest: The author declares that the content of this article has no conflict of interest. The author has no relevant affiliations or financial involvement with any organization or entity with a financial interest in or financial conflict with the subject matter or materials discussed in this article.

References

1. Bhalla, N.; Jolly, P.; Formisano, N.; Estrela, P. Introduction to biosensors. *Essays Biochem.* **2016**, *60*, 1–8. [CrossRef] [PubMed]
2. Esmaeili, C.; Abdi, M.M.; Mathew, A.P.; Jonoobi, M.; Oksman, K.; Rezayi, M. Synergy Effect of Nanocrystalline Cellulose for the Biosensing Detection of Glucose. *Sensors* **2015**, *15*, 24681–24697. [CrossRef] [PubMed]
3. Montañez, M.I.; Hed, Y.; Utsel, S.; Ropponen, J.; Malmström, E.; Wågberg, L.; Hult, A.; Malkoch, M. Bifunctional Dendronized Cellulose Surfaces as Biosensors. *Biomacromolecules* **2011**, *12*, 2114–2125. [CrossRef] [PubMed]
4. Fatimi, A.; Axelos, M.A.V.; Tassin, J.F.; Weiss, P. Rheological Characterization of Self-Hardening Hydrogel for Tissue Engineering Applications: Gel Point Determination and Viscoelastic Properties. *Macromol. Symp.* **2008**, *266*, 12–16. [CrossRef]
5. Fatimi, A.; Tassin, J.F.; Quillard, S.; Axelos, M.A.; Weiss, P. The rheological properties of silated hydroxypropylmethylcellulose tissue engineering matrices. *Biomaterials* **2008**, *29*, 533–543. [CrossRef] [PubMed]
6. Fatimi, A.; Tassin, J.F.; Turczyn, R.; Axelos, M.A.; Weiss, P. Gelation studies of a cellulose-based biohydrogel: The influence of pH, temperature and sterilization. *Acta Biomater.* **2009**, *5*, 3423–3432. [CrossRef] [PubMed]
7. Kamel, S.A.; Khattab, T. Recent Advances in Cellulose-Based Biosensors for Medical Diagnosis. *Biosensors* **2020**, *10*, 67. [CrossRef] [PubMed]
8. Deshpande, S.D.; Kim, J.; Yun, S.-R. Studies on conducting polymer electroactive paper actuators: Effect of humidity and electrode thickness. *Smart Mater. Struct.* **2005**, *14*, 876–880. [CrossRef]

9. Godfrey, R.E. Polymer-Coated Optical Structures and Methods of Making and Using the Same. Granted Patent US 4992385 A; Washington, DC, USA, 12 February 1991.
10. World Intellectual Property Organization. The Patentscope. Available online: https://patentscope.wipo.int (accessed on 2 December 2021).
11. World Intellectual Property Organization. Patentscope Fields Definition. Available online: https://patentscope.wipo.int/search/en/help/fieldsHelp.jsf (accessed on 2 December 2021).
12. Cambia Institute. The Lens Patent Data Set. Version 8.0.14. Available online: https://www.lens.org (accessed on 2 September 2021).
13. European Patent Office. Espacenet Glossary. Version 1.24.1. Available online: https://worldwide.espacenet.com/patent (accessed on 2 December 2021).
14. World Intellectual Property Organization. IPC Publication. IPCPUB v8.5. Available online: https://www.wipo.int/classifications/ipc/ipcpub (accessed on 2 December 2021).
15. World Intellectual Property Organization. Guide to the International Patent Classification (IPC). Available online: https://www.wipo.int/edocs/pubdocs/en/wipo_guide_ipc_2020.pdf (accessed on 2 December 2021).
16. Bloomberg. i-Sens Company Profile. Available online: https://www.bloomberg.com/profile/company/099190:KS (accessed on 5 January 2022).
17. Bloomberg. Osang Healthcare Company Profile. Available online: https://www.bloomberg.com/profile/company/036220:KS (accessed on 5 January 2022).
18. World Intellectual Property Organization. What is Intellectual Property? Frequently Asked Questions: Patents. Available online: https://www.wipo.int/patents/en/faq_patents.html (accessed on 2 September 2021).
19. United States Patent and Trademark Office. Manual of Patent Examining Procedure: Ownership/Assignability of Patents and Applications. Available online: https://www.uspto.gov/web/offices/pac/mpep/mpep-0300.html (accessed on 13 December 2021).
20. Fatimi, A. Patent landscape analysis of hydrogel-based bioinks for 3D bioprinting. *Mater. Proc.* **2021**, *7*, 3. [CrossRef]
21. Fatimi, A. Patentability of biopolymer-based hydrogels. *Mater. Proc.* **2021**, *3*. [CrossRef]
22. Baker, C.O.; Behrenbruch, C.; Rahib, L. Functionalized Polymer Biosensor. Patent WO 2010124270 A1; PCT, Seattle, WA, USA, 28 October 2010.
23. Kim, J.H. Disposable Biosensors Made with Tin Oxide—Cellulose Nanocomposite, Manufacturing Method of the Same and Measuring Method Using the Same. Patent KR 20120050359 A; Korea, 18 May 2012.
24. Wilsey, C.D. Determination of Blood Glucose in a Small Volume and in a Short Test Time Using a Chemical Coating Including Binders and Very Short Read Potentials. Granted Patent US 8349168 B2; USA, 11 December 2012.
25. Jeon, S.-I.; Jung, S.-K.; Im, S.-M.; Lee, E.-S.; Kim, S.-H.; Suk, H.-S.; Kim, W.-D.; Koh, J.-M. Reagent Composition for Biosensor and Biosensor Having the Same. Patent EP 2821497 A2; Europe, 7 January 2015.
26. Agarwal, A.K.; Goel, V.K.; Kim, D.-S.; Lin, B. A Biosensor Device to Detect Target Analytes In Situ, In Vivo, and/or in Real Time, and Methods of Making and Using the Same. Patent US 20180271423 A1; Toledo, OH, USA, 27 September 2018.
27. Pu, Z.; Li, C.; Li, D. Flexible Epidermal Electrochemical Biosensor Based on Conductive Hydrogel. Patent CN 111839532 A; Tianjin, China, 30 October 2020.

Proceeding Paper

Monolithically Integrated Label-Free Optical Immunosensors [†]

Panagiota Petrou [1,*], Eleni Makarona [2], Ioannis Raptis [2], Sotirios Kakabakos [1] and Konstantinos Misiakos [2]

1   Immunoassays/Immunosensors Lab, Institute of Nuclear & Radiological Sciences & Technology,
    Energy & Safety, National Centre for Scientific Research "Demokritos", 15341 Aghia Paraskevi, Greece;
    skakab@rrp.demokritos.gr
2   Institute of Nanoscience and Nanotechnology, National Centre for Scientific Research "Demokritos",
    15341 Aghia Paraskevi, Greece; e.makarona@inn.demokritos.gr (E.M.); i.raptis@inn.demokritos.gr (I.R.);
    k.misiakos@inn.demokritos.gr (K.M.)
*   Correspondence: ypetrou@rrp.demokritos.gr; Tel.: +30-2106503819
†   Presented at the 2nd International Electronic Conference on Biosensors, 14–18 February 2022; Available online:
    https://sciforum.net/event/IECB2022.

Abstract: Amongst label-free optical sensors, those relying on silicon photonics are especially promising for the development of small-sized devices appropriate for applications at the point-of-need. In this context, our work over the last 10 years has focused on the development of silicon photonic chips that combine all optical components, both active and passive, onto the same substrate. The approach followed for this monolithic integration, as well as the application of the different silicon photonic chip versions as immunosensors for the determination of single or panels of analytes, related to biodiagnostics or the food safety sector, will be presented.

Keywords: optical sensors; label-free detection; monolithic integration; biodiagnostics; food analysis

Citation: Petrou, P.; Makarona, E.; Raptis, I.; Kakabakos, S.; Misiakos, K. Monolithically Integrated Label-Free Optical Immunosensors. *Eng. Proc.* **2022**, *16*, 11. https://doi.org/10.3390/IECB2022-12283

Academic Editors: Giovanna Marrazza and Sara Tombelli

Published: 15 February 2022

Publisher's Note: MDPI stays neutral with regard to jurisdictional claims in published maps and institutional affiliations.

Copyright: © 2022 by the authors. Licensee MDPI, Basel, Switzerland. This article is an open access article distributed under the terms and conditions of the Creative Commons Attribution (CC BY) license (https://creativecommons.org/licenses/by/4.0/).

1. Introduction

Biosensors have been the subject of intensive research effort aiming to overcome the limitations of classical analytical systems and provide solutions for on-site determinations for more than 50 years, since the first report of Clark and Lyons [1]. As a result, an enormous variety of biosensing principles has been developed, most of which fall into one of the following categories: mass-sensitive, electrochemical, or optical sensors.

Optical biosensors are less vulnerable compared to mass-sensitive and electrochemical ones to interferences from sample components, parasitic electrical signals, and fluctuations in temperature, and therefore more suitable for point-of-need applications. In addition, with respect to electrochemical ones, optical biosensors have the inherent advantage of reduced crossover signals, enabling their application to the simultaneous detection of multiple analytes (multiplexed determinations), which is of high value for fields such as biodiagnostics and food analysis [2–4].

Independently of the transduction principle, biosensors can be also divided into two categories depending on whether or not they implement labels for analyte determination. Despite the fact that optical biosensors employing labels are considered, in general, more sensitive than the label-free ones [5], they are more suited for laboratory rather than point-of-need applications due to the size and complexity of the relative instrumentation. On the other hand, label-free biosensors have also been improved, especially regarding their analytical performance and detection sensitivity, which reaches in many cases that of biosensors employing labels [2–4], as result of the progress made in the last two decades in the field of nanotechnology.

In general, label-free sensors produce a signal when the analyte binds to the specific recognition molecule that has been immobilized onto the transducer surface, thus enabling the real-time monitoring of the binding reaction. This feature allows the performance of kinetic measurements and, usually, leads to faster assays than the label-based sensing methods. The optical label-free methods can be divided into two main categories, reflectometric and refractometric [6].

In reflectometric sensors, the transducer consists of layers with different refractive index, to which a layer of dielectric material and the layer of recognition molecules are included. The most common reflectometric sensing method is the one known as reflectometric interference spectroscopy (RIfS), introduced by Gauglitz et al. in 1991 [7]. The RIfS transducer is, usually, a glass slide modified with a thin layer of transparent dielectric material (e.g., SiO_2, SiO_2-Ta_2O_5) on top of which the specific recognition molecules are immobilized. When white light strikes the glass substrate from the backside, the partial beams are reflected at each interface, interfere, and create a reflectance spectrum with alternating maxima and minima corresponding to constructive and destructive interference of the reflected radiation. Binding reactions taking place on top of the dielectric layer increase the optical path length, causing a shift in the reflectance spectrum, which is proportional to the thickness increase and consequently to the concentration of the reacting molecules. Over the years, several variations of the initial biosensing method have been presented, in which the white light source and whole reflection spectrum recording were replaced by light sources with a narrow spectral width, enabling multiplex detection in microtiter plates or sensor arrays [8,9], as well as single-wavelength light set-ups suitable for imaging [10]. Moreover, apart from glass, other materials have been investigated as substrates, including porous silicon as is or in combination with dielectric materials [11,12]. White Light Reflectance Spectroscopy (WLRS) is another label-free reflectometric detection method based on the reflection of a white light beam from a silicon chip surface with a dielectric layer on top that is has been engineered so that the reflected spectrum to have at least an interference fringe in the visible spectrum [13]. This sensing approach has been applied over the years for the determination of either single or multiple analytes, related to human disease diagnostics or food safety, using the same optical set-up; the latter was possible through appropriate engineering of the chip [14]. Moreover, effort was devoted to the automation of the assay and signal processing, aiming towards the development of a small-sized device for on-site determinations [15].

Refractometric transducers rely on another transduction principle, that of evanescent field optics. The evanescent field is generated by the electromagnetic field of the light as it transverses a waveguide by means of total internal reflection. The main characteristic of this field is that it extends to a depth ranging from a few tens to a few hundreds of nanometers, depending on the waveguide material and geometry, and its intensity is exponentially reduced as the distance from the waveguide increases. Nonetheless, the evanescent field is very sensitive to changes in the refractive index on the waveguide surface and can thus "sense" the interaction between molecules immobilized onto the waveguide surface with their counterpart molecules. The change in the thickness of the biomolecular layer influences the evanescent field and its coupling back into the waveguide, causing a change in the intensity, polarity, or phase of the waveguided light. The major categories of refractometric sensors include surface plasmon resonance (SPR), grating coupler, photonic crystal, ring resonator, and interferometric sensors (Figure 1).

Surface plasmon resonance (SPR) is the most widely explored label-free optical detection principle [16]. SPR biosensors are based on the immobilization of the recognition molecule on top of a gold layer deposited on a prism, a grating coupler, or a dielectric waveguide. Light that passes from these components and strikes the gold layer at a certain angle can excite the free conducting electrons (plasmons) of the metal and create a surface plasmon wave at the solution/gold interface. This wave is very sensitive to refractive index changes at the gold layer surface, and as a result, the angle of incident light has to change during the course of a binding reaction to preserve the surface plasmon wave. Thus, it is possible to monitor in real time binding reactions, by monitoring the incident light resonance angle. Despite the fact that SPR has found numerous applications in diverse fields and a few companies have commercialized devices based on this transduction principle, the need for external optical components has restricted their miniaturization and confined the use of SPR instruments in the lab. A way to surpass these limitations is offered by the localized SPR or LSPR transduction approach, in which the continuous metal layer has been replaced by metallic nanostructures (nanospheres, nanorods, or nanodisks) of sub-wavelength size [2]. The light that strikes the nanostructures excites the metal free electrons and when resonance is achieved, certain wavelengths are scattered from the nanostructures. Thus, binding reactions can be monitored in real time as shifts in the resonance wavelength. Regarding the question about which format, the classical SPR or LSPR, is more sensitive, the literature reports show that SRP might be more sensitive than LSPR in terms of bulk refractive index changes, while LSPR might be equally sensitive to changes that occur in close proximity to the surface, and therefore more suitable for the monitoring of biomolecular interactions [17].

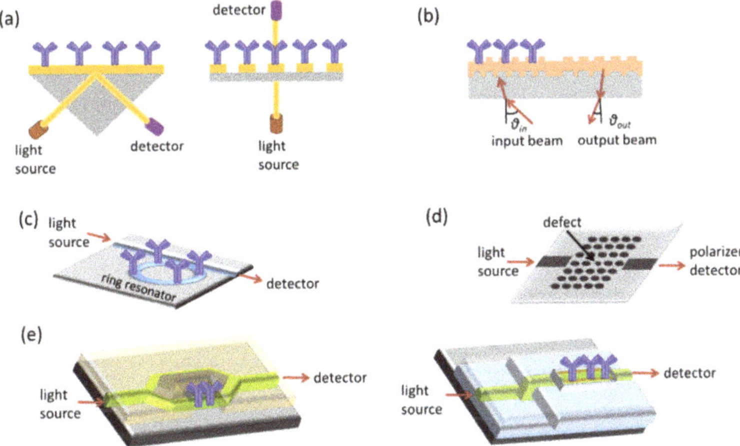

Figure 1. Schematic presentation of label-free optical sensing methods: (**a**) SPR, (**b**) grating couplers, (**c**) ring resonators, (**d**) photonic crystals, and (**e**) interferometers.

Nevertheless, the analytical performance is one only of the parameters that should be taken into consideration when the application of a sensing system outside a laboratory environment is sought [18]. To this end, the simplicity of the measurement procedure (ideally, the user should only load the sample), the cost-effectiveness in terms of both the instrumentation and the consumables required, and the ability to work with complex matrices, but, most importantly, the potential for miniaturization are of upmost importance. Thus, sensors based on integrated optics can be more easily downsized, integrated with external optical components, and interconnected with fluidic and electronic modules, and therefore are the most promising candidates for incorporation into portable devices. Other advantages of integrated optical sensors are the high versatility of materials and technologies available for their fabrication and the ability to create arrays of sensors suitable for multiplex determinations [19]. Although polymers have been implemented as substrates for the fabrication of integrated optical devices, silicon remains the material of choice for high-performance optical devices. A short description of the detection principles of the different integrated optical sensor categories is provided below.

2. Integrated Optical Sensors

In grating coupler sensors, the transducer is a planar waveguide with a grating on its surface, i.e., a periodic pattern, to enable light in-coupling and transmission along the waveguide. The light coupling is sensitive to changes in the refractive index of the medium over the waveguide, allowing monitoring of binding reactions occurring onto the waveguide surface by monitoring the incident light in-coupling angle. Moreover, the interaction of the biomolecular layer with the waveguided photons through the evanescent wave field makes it possible to monitor its building-up by monitoring changes in the light out-coupling angle. This latter configuration is advantageous compared to the first one because there is no need for precision alignment of the light to the grating, thus leading to simpler experimental set-ups [20]. Moreover, several improvements regarding the light in- and out-coupling onto the waveguide allowed multi-analyte determinations using arrays of grating coupler-based sensors [21].

Ring resonators combine a linear waveguide with a circular one in which the light propagating through the linear one is coupled through the evanescent wave field and propagates in the form of whispering-gallery modes. Changes in the refractive index on the ring surface change the spectral position of the whispering-gallery modes and thus resonance is achieved at a different wavelength of the incident light. The fact that the light propagating in the ring interacts multiple times with the molecules on its surface makes ring resonators considerably more sensitive than linear waveguides of the same length. Sensitivity is also enhanced when, instead of a 2D format (microdisk or microring), the resonator acquires a 3D format (microtoroid) [22].

Another category of integrated optical sensors is those based on photonic crystals, i.e., periodical nanostructures on a crystal, which allows the propagation of selected wavelengths of the incident light. If a "defect" is introduced in the periodic structure, the propagation of light is accomplished only when resonance is achieved. As this resonance is influenced by refractive index changes around the defect area, label-free monitoring of bioreactions taking place on the photonic crystal surface is possible. The most popular photonic crystal sensor configuration is a waveguide with arrays of holes arranged in lines or hexagonal lattices on which the defect is created either by missing holes in the pattern or by changing the spacing or size of the holes at some point of the waveguide [23]. In general, photonic crystal sensors are less sensitive compared to other types of integrated optical transducers.

Integrated interferometers could attain several configurations, the most popular of which are Mach–Zehnder (MZI), Young (YI), or bimodal interferometers (BI). Integrated MZIs are linear waveguides that at some point split into two arms, the sensing and the reference, that recombine to a single waveguide after a certain distance. The whole structure is covered by a cladding layer, except from a part of the sensing arm, the sensing window, that is modified with the binding molecules. When a biomolecular interaction takes place on the sensing window, the associated change in the refractive index induces a phase difference between the light that propagates into the sensing arm to that propagating into the reference ones. Thus, the output light intensity has a cosine dependence to the input light. Consequently, MZIs are more sensitive when they operate away from the extrema of the interference spectrum, where the "sensitivity" in refractive index changes is almost negligible. The analytical performance of sensors based on MZIs depends on the sensing arm interaction length, the geometrical characteristics of the waveguide (symmetric MZI with equal lengths of the sensing and the reference arm or asymmetric), and the difference in the refractive index of the waveguide (e.g., glass, SiO_2, Si_3N_4, polymer) and cladding layer material (usually SiO_2) [24,25]. The majority of MZI-based instruments employ monochromatic light sources, i.e., lasers; nonetheless, to reduce the size and complexity of instrumentation, broadband light sources have been also explored instead of lasers. These light sources can be external [26] or integrated to the same substrate with the transducers [27]. The latter will be further discussed in the following section.

Young interferometers are also comprise a waveguide that splits into two arms, which, however, in contrast to MZIs, do not converge on the chip but the two light beams interfere in free space, creating an "interferogram" that can be depicted on a CCD camera [28]. Although there is theoretical and experimental evidence that YIs can be more sensitive for a particular application than sensors based on SPR, grating couplers, or reflectometric interference spectroscopy (RIfS) [28], there are very few reports of biosensors based on YIs in the literature compared to other integrated optical transducers.

Bimodal interferometers represent a category of interferometric sensors that relies on a single waveguide with two different light propagation zones. The first zone supports a single mode and the second two modes (zero- and first-order modes). This means that refractive index changes occurring at the waveguide surface, i.e., due to a biomolecular reaction, effect a change in the interference pattern recorded at the waveguide output since the propagation velocity of two modes depends on the refractive index of the medium on top of the waveguide [29]. Bimodal interferometers have comparable analytical performance to single-wavelength MZIs and in addition have the potential of multiplexed determinations and integration into small-sized instruments [29].

Prior to delving further into monolithically integrated optical sensors, the principles of immunochemical detection methods will be briefly discussed.

3. Principles of Immunochemical Detection Methods

The widespread use of antibodies as recognition elements in biosensors results from their high selectivity and sensitivity but also from the fact they have been used for years in standard immunochemical detection methods, as are the enzyme immunoassays performed on microtiter plates (i.e., ELISAs). Thus, although, in principle, some optical transduction techniques have the necessary sensitivity to monitor directly the antigen–antibody binding, in many cases, this ability is limited to high-molecular-weight analytes, the binding of which can cause a substantial change in sensor response. Therefore, the assay formats usually applied in microtiter plate immunoassays are also employed in immunosensors, namely the competitive and the non-competitive assay format (Figure 2).

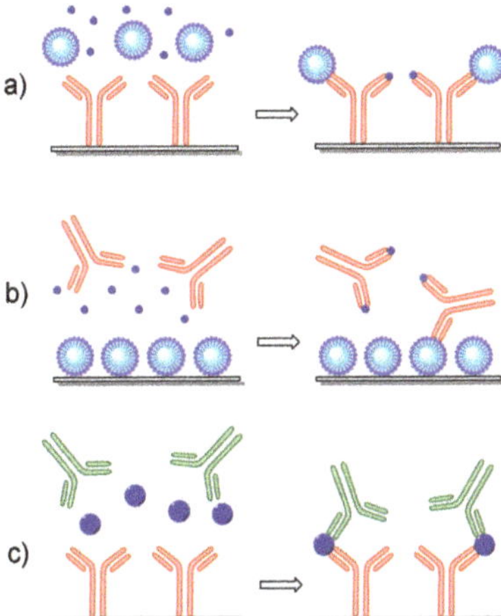

Figure 2. Schematic presentation of different assay formats applied to optical immunosensors: (**a**) a competitive immunoassay with the antibody immobilized onto the transducer, (**b**) a competitive immunoassay with the antigen immobilized onto the transducer in form of a protein conjugate, and (**c**) a non-competitive immunoassay employing a pair of antibodies, one immobilized onto the transducer and the other for detection of analyte molecules bound onto the immobilized antibody.

Non-competitive immunoassays are better suited for low-molecular-weight analytes such as toxins, pesticides, antibiotics, pharmaceutical residues, etc. There are two general approaches based on immobilization onto the transducer surface of either the antigen-specific antibody (Figure 2a) or the antigen itself (Figure 2b). In the first approach, the antigen concentration in the samples is determined through its competition with an antigen bound onto a carrier (usually selected to enhance the sensor signal) for coverage to the immobilized antibody binding sites. In the second approach, the transducer is modified with the antigen either directly or in the form of an antigen–protein conjugate. Again, determination of the antigen concentration in a sample is realized through its competition with the surface-bound antigen for coverage of the antibody binding sites. In both approaches, the signal recorded in the transducer is inversely proportional to the antigen concentration in the sample; thus, the highest signal is received in the absence of the antigen (zero calibrator). Both of the above-described approaches can be applied for an antigen–antibody pair; nonetheless, the second might be preferable when the stability of the immobilized biomolecule is considered since low-molecular-weight antigens or their protein conjugates are less prone to loss of functionality compared to antibodies.

The non-competitive or sandwich immunoassay format is usually followed when the antigen has at least two antigenic determinants or epitopes in its molecule since it requires the combination of two antibodies that do not compete with each other for binding to the antigen. Thus, to perform a non-competitive immunoassay (Figure 2c), one of the antibodies should be immobilized onto the transducer surface (capture antibody) and a second one (detection or reporter antibody) is attached on a different epitope of the surface-bound antigen molecules, forming a "sandwich". In sensors based on label-free transduction principles, the detection antibody might be non-labeled; however, the use of labeled antibodies is possible even in these cases, aiming at the enhancement of the signal derived by the binding of the detection antibody, and consequently the analytical sensitivity of the assay, i.e., the lowest antigen concentration that can be determined.

4. Monolithically Integrated Interferometric Transducers

As discussed in Section 2, the majority of the integrated optical sensors, including the interferometric ones, relied on external monochromatic light sources, i.e., lasers. Due to the low light coupling efficiency to the waveguides, lasers with high intensity of emitted light had to be implemented, which were bulky and energy-demanding. Thus, it was almost impossible to build portable systems based on these components. To reduce the instrumentation size, broadband light sources were coupled to integrated MZIs in an attempt to realize sensors suitable for on-site determinations [30,31]. An external spectrum analyzer working in the 1200–1700 nm spectral region was employed to record the output light. Although the detection sensitivities achieved with these sensors were lower compared to those implementing monochromatic light sources, the use of a broadband light source solved the phase ambiguity issue of monochromatic MZI, which arises from the fact that information might be lost if the refractive index changes resulted in phase shifts equal to or multiple-times equal to 2π. Introduction of an on-chip optical spectral analyzer along with an array of MZIs and a grating for broadband light in-coupling improved the sensor performance [26,32]. Thus, a limit of detection (LOD) of 6×10^{-6} RIU in terms of refractive index was achieved, which was comparable to the LODs of monochromatic MZIs. Broadband light sources have been also combined with integrated Young interferometers, for which it was predicted by simulations that the implementation of these light sources would enable discrimination between refractive index changes caused by binding reactions rather than medium changes [33]. These theoretical predictions have not been proved experimentally so far.

In all efforts mentioned, the light source, either monochromatic or broadband, was external, and additional optical components, lenses, gratings, etc., were required to couple the light onto the integrated waveguides. The way to extract light from silicon has been known since the 1960s [34]; however, the low intensity of these light sources, combined with the difficulty to efficiently couple this light to waveguides made onto the same substrate, prohibited their implementation in biosensing devices for almost 50 years. Then, a new approach to monolithically integrate silicon waveguides, light sources, and detectors into the same silicon substrate was introduced [35]. The integrated light source was a silicon avalanche diode (LED), which emitted light covering all the visible and near-infrared spectrum when reversed-biased beyond its breakdown point. The realization of working sensors based on silicon components all fabricated onto the same substrate required inventive solutions of two problems, the bending of the waveguides towards the

LED and the photodetector to avoid light losses, and the alignment of the LED with the waveguide. The process followed to solve these two problems is summarized in Figure 3. Thus, at first, a 2-μm-thick thermally grown silicon dioxide layer was deposited on the silicon substrate, on which the positions of the LED and photodetector were defined through photolithography and etching of the silicon dioxide layer. To these openings, the base of the avalanche junction (N+) and the p/n junction at the photodetector side were formed by phosphorus implantation (Figure 3i). Then, an additional 2-μm-thick silicon dioxide layer was created by deposition of tetraethyl orthosilicate (TEOS) (Figure 3ii). The TEOS layer was etched in CHF_3 plasma to create curvatures towards the LED and photodetector side (Figure 3iii). Low-pressure chemical vapor deposition (LPCVD) of a silicon nitride film with a thickness of around 150 nm was then performed using a mixture of NH_3 and SiH_2Cl_2 (Figure 3iv). Photolithography and etching in CHF_3 plasma were applied again to create the strip waveguides, one end of which was just above the avalanche diode base area. Following this, the avalanche diode was created by boron implantation through the nitride film, which acted as a mask guiding the formation of the LED immediately under the up-going segment of the waveguide (Figure 3v). Finally, the silicon dioxide cladding layer was formed by deposition, and etched above the waveguides to define the sensing windows. This process resulted in less than 1-micron misalignment of the LED with respect to waveguide and a light coupling efficiency that reached almost 40% [35].

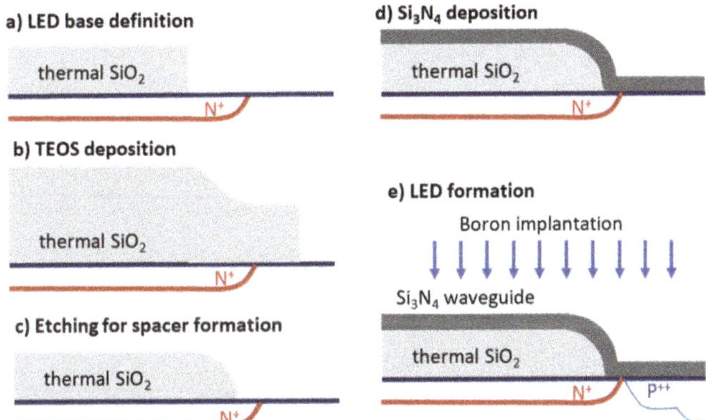

Figure 3. Schematic of the process followed for integrated transducer fabrication: (**a**) formation of avalanche diode base, (**b**) silicon dioxide deposition, (**c**) plasma etching for bending spacer creation (**d**), deposition of silicon nitride film and waveguide patterning, and (**e**) boron implantation through the silicon nitride film for self-aligned formation of the LED.

In the first version of monolithically integrated silicon optical sensors, the LEDs were coupled to linear silicon nitride waveguides and signal transduction was based on monitoring the drop in the intensity of light transmitted through the waveguide and recorded by the photodetector due to its interaction with light-absorbing labels attached to targeted analytes [35,36]. Nonetheless, shortly after, the linear waveguides were replaced by MZIs, ten of which were arranged in a fan-like way on a single chip, each one connected to its own integrated LED and all of the them converging to the same output point [27,37,38]. Moreover, to ensure that the light transmitted was monomodal, a mode filter was introduced prior to the waveguide. It was theoretically and experimentally shown using these chips that broadband Mach–Zehnder interferometry could surpass both the phase ambiguity and signal fading of standard single-wavelength MZIs [27,37,38]. This was ascribed to the fact that a given change in the refractive index of the medium above the sensing waveguide window corresponds to a different phase shift for each wavelength, and thus minute changes in the refractive index could be detected more accurately by processing the full transmission spectrum rather than monitoring a single wavelength.

Two configurations of the integrated MZIs have been realized, a fully (Figure 4a) and a semi-integrated one (Figure 4b). The former incorporated on a single chip, along with the integrated LEDs and the MZIs, as a detector, an integrated photodiode and had an LOD in terms of refractive index of 1×10^{-5} RIU. This configuration was applied to monitor binding reactions, including the interaction of immobilized biotin with streptavidin and mouse IgG with an anti-mouse IgG antibody, with LODs of 1 nM and 10 nM for streptavidin and anti-mouse IgG, respectively [37]. Despite the promising analytical results, it was obvious that the fully integrated version suffered from the limitation of phase ambiguity and signal fading of the monochromatic MZIs since the use of a photodetector for signal collection suppressed the information available in the spectrum to a single value. Thus, the semi-integrated version that employed an external spectrophotometer for recording the transmission spectrum was further exploited as an immunosensor [38–44]. The spectrum recorded contained two characteristic frequencies, one for the TE and one for the TM mode, which could be discriminated through Fourier transform, offering the ability to monitor binding events by monitoring either of them without significant effects on the analytical performance. The analytical sensitivity of the semi-integrated device, with respect to the bulk refractive index, was determined at 5×10^{-6} RIU. For the binding assays of biotinylated bovine serum albumin with streptavidin and anti-mouse IgG with mouse IgG, LODs of 5 pM and 32 pM, respectively, were determined, demonstrating the considerably higher detection sensitivity of the semi-integrated configuration as compared to the fully integrated one [38].

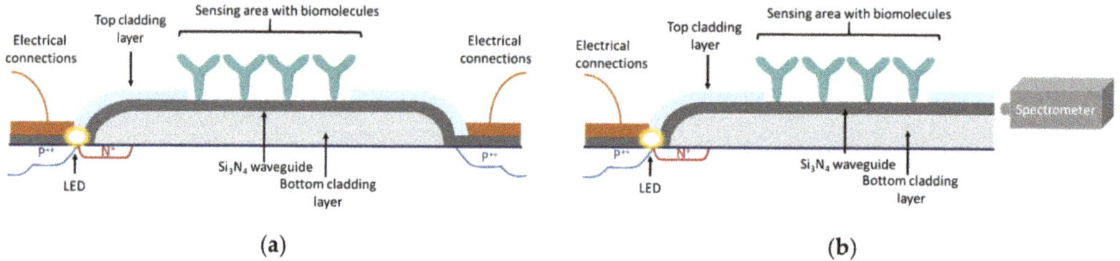

Figure 4. Cross-section depiction of fully integrated (**a**) and the semi-integrated (**b**) broadband MZI configuration.

Thus, the semi-integrated configuration was implemented in both single- and multi-analyte immunochemical determinations. The first application related to the detection of goat milk adulteration with bovine milk through the immunochemical determination of bovine k-casein [39]. A competitive immunoassay was developed using MZIs modified with bovine k-casein and a specific antibody against this protein. The LOD of the assay was 0.04% (v/v) bovine in goat milk and the assay's dynamic range was from 0.1 to 1.0% (v/v). Moreover, the assay was completed in 10 min. The same assay format and reagents were applied to detect mozzarella or feta adulteration with bovine cheese [40]. In this case also, the assay duration was less than 10 min and the limit of quantification for bovine cheese in mozzarella and feta cheese was 0.5 and 0.25% (w/w), respectively—well below the maximum allowable content of bovine milk in mozzarella and feta (1% w/w) according to the EU regulations. The mycotoxin ochratoxin A was also detected in beer samples following a competitive immunoassay format. The assay detection limit was 2.0 ng/mL and its dynamic range 4.0–100 ng/mL [41]. The high detection sensitivity of integrated MZI-based immunosensors allowed the determination of C-reactive protein (CRP), a widely used inflammation marker, in human serum samples in approximately 5 min by monitoring the direct binding of the analyte to an antibody immobilized onto the windows of the MZIs' sensing arms [42]. A quantification limit of 4.2 ng/mL was achieved, allowing an at least 100-times dilution of the serum that alleviated any matrix effect onto the sensor response. In addition, the CRP concentrations determined in human serum samples with the sensor were in excellent agreement with those determined for the same samples by a clinical analyzer, supporting the accuracy of the measurements performed with the sensor.

The realization of arrays of 10 integrated MZIs on a single chip (with dimensions $4.25 \times 8.0 = 34$ cm^2) offered the possibility for multiplexed determinations through modification of the sensing arms' window areas of different MZIs with different binding molecules. Thus, by immobilizing protein conjugates of three mycotoxins, each one on three MZIs of the chip, and leaving the last one for the determination of the non-specific binding signal, the simultaneous determination of aflatoxin B1, fumonisin B1, and deoxynivalenol in beer samples was accomplished following a competitive immunoassay format. The assays' LODs were 0.8, 5.6, and 24 ng/mL for aflatoxin B1, fumonisin B1, and deoxynivalenol, respectively, and the overall assay duration was only 12 min [43]. Analysis of different types of beers produced around the world showed that the sensor could perform as well as established instrumental laboratory methods [44]. Following a similar procedure, the simultaneous detection of four allergens, bovine milk protein, peanut protein, soy protein, and gliadin, was demonstrated [45]. In this case, the four allergenic proteins were immobilized onto the integrated MZIs to perform a competitive immunoassay that was completed in 6.5 min, providing LODs of 0.04, 1.0, 0.80, and 0.10 µg/mL, for bovine milk protein, peanut protein, soy protein, and gliadin, respectively. Evaluation of sensor performance through the analysis of samples from the cleaning-in-place system of a dairy facility showed that the results obtained with the sensor were in good agreement with those acquired by commercial ELISAs for the determination of each antigen separately.

The use of integrated silicon LEDs as light sources for interferometric sensors was expanded to integrated YIs [46]. Similarly to MZIs, it was shown that the implementation of a polychromatic light source led to an interferogram consisting of two distinct fringe packets, one for each polarization, making feasible the independent determination of the phase signal for the two polarizations. The sensor was evaluated using the binding reactions of streptavidin and anti-mouse IgG antibody with immobilized biotin and mouse IgG, respectively. Using a concentration of streptavidin of 1 nM and anti-rabbit IgG of 10 nM, peak shifts of more than one period were observed, indicating that LODs down to the sub-nM range could be achieved [47]. Nonetheless, the preliminary experiments revealed also the weak points of the set-up and especially the need for precise alignment of the chip with respect to the CCD array used to record the signal. Therefore, further advancements regarding the experimental set-up are required in order to benefit in full from the analytical performance of the integrated YIs.

The need for external detectors, especially in the case of integrated MZIs where a spectrophotometer was required, compromised the portability potential offered by the small chip size. Thus, a spectrum analyzer and a photodiode array were integrated into the chip with the 10 integrated MZIs and the ultimate degree of integration of all passive and active components on a single chip was achieved without increasing considerably the overall chip size (37 mm^2). Based on these chips, a portable instrument was manufactured that encompassed in a $21 \times 17 \times 7$ cm^3 case the chip docking station, a micropump for reagent circulation, and the electronics for turning on the light sources sequentially and recording the signal from the 10 MZIs. The sensor provided an LOD of 60 pM for a direct binding assay of anti-mouse IgG antibody onto immobilized mouse IgG and 8 pM for a non-competitive immunoassay of C-reactive protein, both completed in around 10 min. The good analytical performance of these integrated sensors and the small chip and instrumentation size are the main pros of this approach, promoting the application of the sensor and instrument at the point-of-need. Regarding the cons of this approach, one can consider the high complexity of chip fabrication and the associated cost, and the need for the more automated handling of the reagents required to perform the assay.

5. Conclusions and Outlook

From the information provided in the previous sections, it is obvious that monolithically integrated optical transducers demonstrate excellent analytical performance in different application fields, ranging from biodiagnostics to environmental monitoring and food safety. Moreover, these transducers seem to provide a viable solution towards the development of portable systems that could be used for analysis outside the analytical laboratory, i.e., at the point-of-need. Towards this goal, an additional advantage of monolithically integrated optical sensors is the fact that they are fabricated with techniques compatible with large-scale production at a reasonable cost. Despite the fact that the research efforts described have resulted in prototypes that work efficiently in a laboratory environment, there are still different aspects to be addressed prior to moving these prototypes to the point-of-need. As mentioned, such a limiting factor is the handling of the sample and, in general, of solutions required for assay performance, which could employ several reaction steps. To this end, a new sensing approach has been exploited in which, contrary to the previous efforts, only the MZIs are integrated onto the chip in such a way that allows the coupling of both the input and output light from the same side of the chip. In the other side of the chip the sensing windows of the MZIs are located, which, after appropriate modification with biomolecules, will allow the monitoring of binding reactions by simple dipping onto solutions, thus abolishing the need for fluidics and fluid manipulation. Provided that the new approach demonstrates the analytical performance of former monolithically integrated optical sensors, it will be the ideal solution for on-site, rapid analytical determinations of high accuracy.

Author Contributions: P.P., E.M., I.R., S.K. and K.M. contributed to the manuscript equally. All authors have read and agreed to the published version of the manuscript.

Funding: This research received no external funding.

Institutional Review Board Statement: Not applicable.

Informed Consent Statement: Not applicable.

Data Availability Statement: No new data were created or analyzed in this study. Data sharing is not applicable to this article.

Conflicts of Interest: The authors declare no conflict of interest.

References

1. Clark, L., Jr.; Lyons, C. Electrode systems for continuous monitoring in cardiovascular surgery. *Ann. N. Y. Acad. Sci.* **1962**, *102*, 29–45. [CrossRef] [PubMed]
2. Vashist, S.K.; Luppa, P.B.; Yeo, L.Y.; Ozcan, A.; Luong, J.H.T. Emerging technologies for next-generation point-of-care testing. *Trend Biotechnol.* **2015**, *33*, 692–705. [CrossRef]
3. Lopez, G.A.; Estevez, M.-C.; Solera, M.; Lechuga, L.M. Recent advances in nanoplasmonic biosensors: Applications and lab-on-a-chip integration. *Nanophotonic* **2017**, *6*, 123–136. [CrossRef]
4. Makarona, E.; Petrou, P.; Kakabakos, S.; Misiakos, K.; Raptis, I. Point-of-Need bioanalytics based on planar optical interferometry. *Biotechnol. Adv.* **2016**, *34*, 209–233. [CrossRef] [PubMed]
5. Walt, D.R. Optical methods for single molecule detection and analysis. *Anal. Chem.* **2013**, *85*, 1258–1263. [CrossRef] [PubMed]
6. Gauglitz, G. Direct optical detection in bioanalysis: An update. *Anal. Bioanal. Chem.* **2010**, *398*, 2363–2372. [CrossRef] [PubMed]
7. Gauglitz, G.; Nahm, W. Observation of spectral interferences for the determination of volume and surface effects of thin films. *Fresenius J. Anal. Chem.* **1991**, *341*, 279–283. [CrossRef]
8. Rothmund, M.; Schütz, A.; Brecht, A.; Gauglitz, G.; Berthel, G.; Gräfe, D. Label free binding assay with spectroscopic detection for pharmaceutical screening. *Fresenius J. Anal. Chem.* **1997**, *359*, 15–22. [CrossRef]
9. Gauglitz, G. Multiple reflectance interference spectroscopy measurements made in parallel for binding studies. *Rev. Sci. Instr.* **2005**, *76*, 062224. [CrossRef]
10. Bleher, O.; Schindler, A.; Yin, M.-X.; Holmes, A.B.; Luppa, P.B.; Gauglitz, G.; Proll, G. Development of a new parallelized, optical biosensor platform for label-free detection of autoimmunity-related antibodies. *Anal. Bioanal. Chem.* **2014**, *406*, 3305–3314. [CrossRef]
11. Arshavsky-Graham, S.; Massad-Ivanir, N.; Segal, E.; Weiss, S. Porous silicon-based photonic biosensors: Current status and emerging applications. *Anal. Chem.* **2019**, *91*, 441–467. [CrossRef] [PubMed]
12. Chen, Y.; Liu, J.; Yang, Z.; Wilkinson, J.S.; Zhou, X. Optical biosensors based on refractometric sensing schemes: A review. *Biosens. Bioelectron.* **2019**, *144*, 111693. [CrossRef] [PubMed]
13. Koukouvinos, G.; Petrou, P.; Goustouridis, D.; Misiakos, K.; Kakabakos, S.; Raptis, I. Development and bioanalytical applications of a white light reflectance spectroscopy label-free sensing platform. *Biosensors* **2017**, *7*, 46. [CrossRef] [PubMed]

14. Anastasiadis, V.; Koukouvinos, G.; Petrou, P.S.; Economou, A.; Dekker, J.; Harjanne, M.; Heimala, P.; Goustouridis, D.; Raptis, I.; Kakabakos, S.E. Multiplexed mycotoxins determination employing white light reflectance spectroscopy and silicon chips with silicon oxide areas of different thickness. *Biosens. Bioelectron.* **2020**, *153*, 112035. [CrossRef]
15. Tsounidi, D.; Koukouvinos, G.; Christianidis, V.; Legaki, E.; Giogli, V.; Panagiotopoulou, K.; Taka, S.; Ekaterinidi, Z.; Kakabakos, S.; Raptis, I.; et al. Development of a point-of-care system based on white light reflectance spectroscopy: Application in CRP determination. *Biosensors* **2021**, *11*, 268. [CrossRef]
16. Yesudasu, V.; Pradhan, H.S.; Pandya, R.J. Recent progress in surface plasmon resonance based sensors: A comprehensive review. *Heliyon* **2021**, *7*, e06321. [CrossRef]
17. Mazzotta, F.; Johnson, T.W.; Dahlin, A.B.; Shaver, J.; Oh, S.-H.; Höök, F. Influence of the evanescent field decay length on the sensitivity of plasmonic nanodisks and nanoholes. *ACS Photonics* **2015**, *2*, 256–262. [CrossRef]
18. Zanchetta, G.; Lanfranco, R.; Giavazzi, F.; Bellini, T.; Buscaglia, M. Emerging applications of label-free optical biosensors. *Nanophotonics* **2017**, *6*, 627–645. [CrossRef]
19. Estevez, M.-C.; Alvarez, M.; Lechuga, L.M. Integrated optical devices for lab-on-a-chip biosensing applications. *Laser Photon. Rev.* **2011**, *6*, 1–25. [CrossRef]
20. Lukosz, W.; Clerc, D.; Nellen, P.M.; Stamm, C.; Weiss, P. Output grating couplers on planar optical waveguides as direct immunosensors. *Biosens. Bioelectron.* **1991**, *6*, 227–232. [CrossRef]
21. Kehl, F.; Etlinger, G.; Gartmann, T.E.; Tscharner, N.S.R.U.; Heub, S.; Follonier, S. Introduction of an angle interrogated, MEMS-based, optical waveguide grating system for label-free biosensing. *Sens. Actuator B* **2016**, *226*, 135–143. [CrossRef]
22. Zhang, X.M.; Choi, H.S.; Armani, A.M. Ultimate quality factor of silica microtoroid resonant cavities. *Appl. Phys. Lett.* **2010**, *96*, 153304. [CrossRef]
23. Dorfner, D.; Zabel, T.; Hurlimann, T.; Hauke, N.; Frandsen, L.; Rant, U.; Abstreiter, G.; Finley, J. Photonic crystal nanostructures for optical biosensing applications. *Biosens. Bioelectron.* **2009**, *24*, 3688–3692. [CrossRef] [PubMed]
24. Sarkar, D.; Gunda, N.S.K.; Jamal, I.; Mitra, S.K. Optical biosensors with an integrated Mach-Zehnder Interferometer for detection of Listeria monocytogenes. *Biomed. Microdev.* **2014**, *16*, 509–520. [CrossRef] [PubMed]
25. Chalyan, T.; Guider, R.; Pasquardini, L.; Zanetti, M.; Falke, F.; Schreuder, E.; Heideman, R.G.; Pederzolli, C.; Pavesi, L. Asymmetric Mach-Zehnder interferometer based biosensors for aflatoxin M1 detection. *Biosensors* **2016**, *6*, 1. [CrossRef] [PubMed]
26. Martens, D.; Ramirez-Priego, P.; Murib, M.S.; Elamin, A.A.; Gonzalez-Guerrero, A.B.; Stehr, M.; Jonas, F.; Anton, B.; Hlawatsch, N.; Soetaert, P.; et al. A low-cost integrated biosensing platform based on SiN nanophotonics for biomarker detection in urine. *Anal. Method.* **2018**, *10*, 3066–3073. [CrossRef]
27. Misiakos, K.; Raptis, I.; Salapatas, A.; Makarona, E.; Botsialas, A.; Hoekman, M.; Stoffer, R.; Jobst, G. Broad-band Mach-Zehnder Interferometers as high performance refractive index sensors: Theory and monolithic implementation. *Opt. Exp.* **2014**, *22*, 8856–8870. [CrossRef]
28. Ymeti, A.; Greve, J.; Lambeck, P.V.; Wijn, R.; Heideman, R.G.; Kanger, J.S. Drift correction in a multichannel integrated optical Young interferometer. *Appl. Opt.* **2005**, *44*, 3409–3412. [CrossRef]
29. Zinoviev, K.E.; Gonzalez-Guerrero, A.B.; Dominguez, C.; Lechuga, L.M. Integrated bimodal waveguide interferometric biosensor for label-free analysis. *J. Lightwave Technol.* **2011**, *29*, 1926–1930. [CrossRef]
30. Luo, D.H.; Levy, R.A.; Hor, Y.F.; Federici, J.F.; Pafchek, R.M. An integrated photonic sensor for in situ monitoring of hazardous organics. *Sens. Actuator B* **2003**, *92*, 121–126. [CrossRef]
31. Li, Y.; Harris, E.; Chen, L.; Bao, X. Application of spectrum differential integration method in an in-line fiber Mach-Zehnder refractive index sensor. *Opt. Express* **2010**, *18*, 8135–8143. [CrossRef] [PubMed]
32. Mulder, H.K.P.; Ymeti, A.; Subramaniam, V.; Kanger, J.S. Size-selective detection in integrated optical interferometric biosensors. *Opt. Exp.* **2012**, *20*, 20934. [CrossRef] [PubMed]
33. Nagata, T.; Namba, T.; Kuroda, Y.; Miyake, K.; Miyamoto, T.; Yokoyama, S.; Miyazaki, S.; Koyanagi, M.; Hirose, M. Single-chip integration of light-emitting diode, waveguide and micromirrors. *Jpn. J. Appl. Phys.* **1995**, *34*, 1282–1285. [CrossRef]
34. Chynoweth, A.G.; McKay, K.G. Photon emission from avalanche breakdown in silicon. *Phys. Rev.* **1956**, *102*, 369–376. [CrossRef]
35. Misiakos, K.; Kakabakos, S.E.; Petrou, P.S.; Ruf, H.H. A monolithic silicon optoelectronic transducer as a real-time affinity biosensor. *Anal. Chem.* **2004**, *76*, 1366–1373. [CrossRef]
36. Misiakos, K.; Petrou, P.S.; Kakabakos, S.E.; Yannoukakos, D.; Contopanagos, H.; Knoll, T.; Velten, T.; DeFazio, M.; Schiavo, L.; Passamano, M.; et al. Fully integrated monolithic optoelectronic transducer for real-time protein and DNA detection: The NEMOSLAB approach. *Biosens. Bioelectron.* **2010**, *26*, 1528–1535. [CrossRef]
37. Misiakos, K.; Raptis, I.; Makarona, E.; Botsialas, A.; Salapatas, A.; Oikonomou, P.; Psarouli, A.; Petrou, P.S.; Kakabakos, S.E.; Tukkiniemi, K.; et al. All-silicon monolithic Mach-Zehnder interferometer as a refractive index and biochemical sensor. *Opt. Exp.* **2014**, *22*, 26803–26813. [CrossRef]
38. Psarouli, A.; Salapatas, A.; Botsialas, A.; Petrou, P.S.; Raptis, I.; Makarona, E.; Jobst, G.; Tukkiniemi, K.; Sopanen, M.; Stoffer, R.; et al. Monolithically integrated broad-band Mach-Zehnder interferometers for highly sensitive label-free detection of biomolecules through dual polarization optics. *Sci. Rep.* **2015**, *5*, 17600. [CrossRef]
39. Angelopoulou, M.; Botsialas, A.; Salapatas, A.; Petrou, P.S.; Haasnoot, W.; Makarona, E.; Jobst, G.; Goustouridis, D.; Siafaka-Kapadai, A.; Raptis, I.; et al. Assessment of goat milk adulteration with a label-free monolithically integrated optoelectronic biosensor. *Anal. Bioanal. Chem.* **2015**, *407*, 3995–4004. [CrossRef]

40. Angelopoulou, M.; Petrou, P.S.; Raptis, I.; Misiakos, K.; Livaniou, E.; Makarona, E.; Kakabakos, S.E. Rapid detection of mozzarella and feta cheese adulteration with cow milk through a silicon photonic immunosensor. *Analyst* **2021**, *146*, 529–537. [CrossRef]
41. Pagkali, V.; Petrou, P.S.; Salapatas, A.; Makarona, E.; Peters, J.; Haasnoot, W.; Jobst, G.; Economou, A.; Misiakos, K.; Raptis, I.; et al. Detection of ochratoxin A in beer samples with a label-free monolithically integrated optoelectronic biosensor. *J. Hazard. Mater.* **2017**, *323*, 75–83. [CrossRef] [PubMed]
42. Psarouli, A.; Botsialas, A.; Salapatas, A.; Stefanitsis, G.; Nikita, D.; Jobst, G.; Chaniotakis, N.; Goustouridis, D.; Makarona, E.; Petrou, P.S.; et al. Fast label-free detection of C-reactive protein using broad-band Mach Zehnder interferometers integrated on silicon chips. *Talanta* **2017**, *165*, 458–465. [CrossRef] [PubMed]
43. Pagkali, V.; Petrou, P.S.; Makarona, E.; Peters, J.; Haasnoot, W.; Jobst, G.; Moser, I.; Gajos, K.; Budkowski, A.; Economou, A.; et al. Simultaneous determination of aflatoxin B1, fumonisin B1 and deoxynivalenol in beer samples with a label-free monolithically integrated optoelectronic biosensor. *J. Hazard. Mater.* **2018**, *359*, 445–453. [CrossRef] [PubMed]
44. Angelopoulou, M.; Petrou, P.S.; Makarona, E.; Haasnoot, W.; Moser, I.; Jobst, G.; Goustouridis, D.; Lees, M.; Kalatzi, K.; Raptis, I.; et al. Ultrafast multiplexed-allergen detection through advanced fluidic design and monolithic interferometric silicon chips. *Anal. Chem.* **2018**, *90*, 9559–9567. [CrossRef]
45. Makarona, E.; Salapatas, A.; Raptis, I.; Petrou, P.; Kakabakos, S.; Stavra, E.; Malainou, A.; Misiakos, K. Broadband Young interferometry for simultaneous dual polarization bioanalytics. *J. Opt. Soc. Am. B Opt. Phys.* **2017**, *34*, 1691–1698. [CrossRef]
46. Savra, E.; Malainou, A.; Salapatas, A.; Botsialas, A.; Petrou, P.; Raptis, I.; Makarona, E.; Kakabakos, S.E.; Misiakos, K. Monolithically-integrated Young interferometers for label-free and multiplexed detection of biomolecules. *Proc. SPIE* **2016**, *9752*, 97520N.
47. Misiakos, K.; Makarona, E.; Hoekman, M.; Fyrogenis, R.; Tukkiniemi, K.; Jobst, G.; Petrou, P.S.; Kakabakos, S.E.; Salapatas, A.; Goustouridis, D.; et al. All-silicon spectrally resolved interferometric circuit for multiplexed diagnostics: A monolithic Lab-on-a-Chip integrating all active and passive components. *ACS Photonic* **2019**, *6*, 1694–1705. [CrossRef]

Proceeding Paper

Electrochemical Detection of Cocaine in Authentic Oral Fluid [†]

Florine Joosten [1,2,*], Marc Parrilla [1,2] and Karolien De Wael [1,2]

[1]   A-Sense Lab, Department of Bioscience Engineering, University of Antwerp, Groenenborgerlaan 171,
       2020 Antwerp, Belgium; marc.parrillapons@uantwerpen.be (M.P.); karolien.dewael@uantwerpen.be (K.D.W.)
[2]   NANOlab Center of Excellence, University of Antwerp, Groenenborgerlaan 171, 2020 Antwerp, Belgium
*   Correspondence: florine.joosten@uantwerpen.be
†   Presented at the 2nd International Electronic Conference on Biosensors, 14–18 February 2022;
     Available online: https://sciforum.net/event/IECB2022.

Abstract: Illicit drug consumption remains a problem to public safety and health, with abuse of illicit drugs having increased significantly over the last years. A concern related to this abuse is driving under the influence of drugs (DUID). Currently, police and law enforcement agencies rely on the use of lateral flow immunoassays (LFAs), which suffer from a lack of specificity. In this report, we present a rapid, sensitive, and affordable electrochemical method for the detection of cocaine in oral fluid (OF) by square-wave adsorptive stripping voltammetry on screen-printed electrodes (SPE). For the first time, the effects of the OF matrix on the electrochemical sensing of cocaine are deeply explored. The interference of endogenous compounds in OF, cutting agents and adulterants is studied. Interestingly, the electrochemical signal for cocaine is shown to be partially suppressed by the biofouling properties of albumin and most probably other proteins present in the OF matrix. Thus, strategies to mitigate these biofouling properties are explored. Subsequently, two sampling methods for OF, expectoration and the use of a commercial OF collection device (i.e., the Intercept i2), are investigated. The developed method shows promising potential in point-of-care testing for recent illicit drug use.

Keywords: square wave voltammetry; oral fluid testing; cocaine; screen printed electrodes; forensic analysis

Citation: Joosten, F.; Parrilla, M.; De Wael, K. Electrochemical Detection of Cocaine in Authentic Oral Fluid. *Eng. Proc.* **2022**, *16*, 13. https://doi.org/10.3390/IECB2022-12284

Academic Editors: Giovanna Marrazza and Sara Tombelli

Published: 15 February 2022

Publisher's Note: MDPI stays neutral with regard to jurisdictional claims in published maps and institutional affiliations.

Copyright: © 2022 by the authors. Licensee MDPI, Basel, Switzerland. This article is an open access article distributed under the terms and conditions of the Creative Commons Attribution (CC BY) license (https://creativecommons.org/licenses/by/4.0/).

1. Introduction

Even during the ongoing COVID-19 pandemic, the consumption of illicit drugs has remained a problem to public health and safety [1]. In the last decade, the number of worldwide drug users has grown at a 30% rate to reach 270 million users in 2018 [2]. In the same year, the highest number of cocaine seizures in Europe was reported [3]. An increasing concern related to the use of illicit drugs and cocaine is that of driving under the influence of drugs (DUID) [4]. In the large-scale European Union (EU) study "Driving under the Influence of Drugs, Alcohol and Medicines (DRUID)" (2012), it has been reported that the detection rate of illicit drugs in the general driving population was 1.9% [5]. This detection rate was higher in seriously injured drivers (2.3–12.6%). The World Health Organization (WHO) has estimated that over 39,600 traffic deaths were caused by DUID in 2013 [4]. Of these deaths, 14% were attributed to the use of cocaine. These numbers show that it is paramount to tackle the DUID issue to improve road safety. A potential solution is to perform more roadside tests to identify and block DUID.

The standard process of illicit drug detection in OF consists of two steps [6]. First, a presumptive test is performed on-site. If the results of this test are positive, they need to be confirmed in the laboratory using techniques such as gas chromatography or liquid chromatography coupled to mass spectrometry (GC- or LC-MS). For presumptive tests, lateral flow immunoassays (LFAs) are currently the gold standard even though they might exhibit some drawbacks as follows: (i) cross-reactivity with similar drugs; (ii) lack of specificity; (iii) time consuming (>5 min); (iv) high-cost; and (v) short shelf lives due to the use of bioreceptors [7,8].

This work aims to tackle the identified issues related to the detection of cocaine in OF by the development of a rapid, affordable, and sensitive sensing method based on electrochemical sensors. The workflow of the sensing method is presented in Figure 1. For the first time, the OF matrix effects on the electrochemical sensing of cocaine are deeply explored by using screen-printed electrodes (SPE). First, the electrochemical behavior of cocaine in buffer solution is investigated by square-wave adsorptive stripping voltammetry (SWAdSV) which adsorption is enabled by the use of a surfactant. Second, the interference of endogenous compounds in OF and cutting agents and adulterants is studied. Interestingly, the electrochemical signal for cocaine is shown to be partially suppressed by the biofouling properties of albumin and most probably other proteins present in the OF matrix. Hence, strategies to mitigate these biofouling properties are explored. Subsequently, two sampling methods for OF, expectoration and the use of a commercial OF collection device (i.e., the Intercept i2), are investigated. Finally, the developed methodology is used to analyze authentic OF spiked with cocaine.

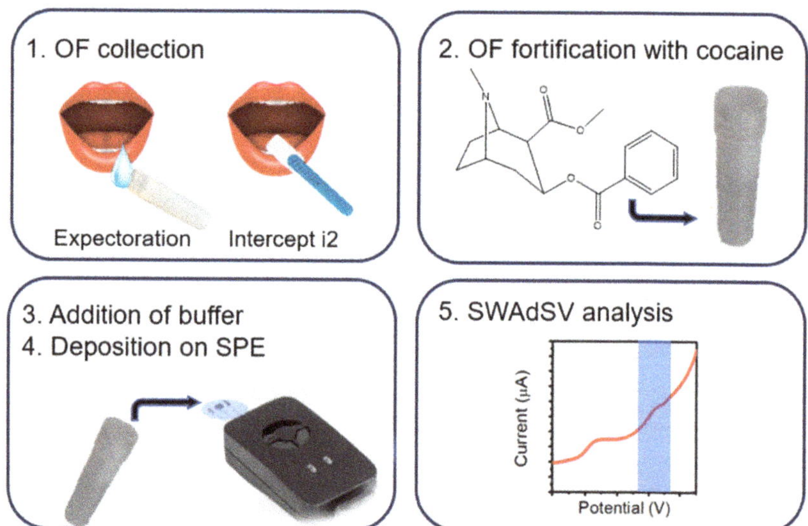

Figure 1. Workflow of the electrochemical sensing method for cocaine in OF. (1) OF collection via expectoration or using the Intercept i2 OF collection device, (2) OF fortification with cocaine, (3) dilution of the OF sample in buffer, (4) deposition of the sample on the electrode, and (5) SWAdSV test and analysis. OF = oral fluid; SPE = screen-printed electrode; SWAdSV = square-wave adsorptive stripping voltammetry.

2. Materials and Methods

2.1. Materials

Standards of cocaine·HCl were purchased from Chiron AS, Norway. Analytical grade salts of potassium chloride, potassium phosphate, sodium phosphate, sodium borate, sodium acetate, potassium hydroxide and sodium hydroxide were purchased from Sigma-Aldrich (Overijse, Belgium). Sodium dodecyl sulphate (SDS) was purchased from Sigma-Aldrich (Overijse, Belgium). Intercept i2 (OraSure Technologies) oral fluid collection devices were purchased from Qarad (Geel, Belgium).

All solutions were prepared in 18.2 MΩ cm^{-1} doubly deionized water (Milli-Q water systems, Merck Millipore, Germany). The pH was measured using a pH-meter (914 pH/Conductometer, 2.914.0020, Metrohm, Switzerland).

2.2. Methods

Electrochemical measurements were performed using a MultiPalmSens 4 (PalmSens, Houten, The Netherlands), a MultiEmstat3 (PalmSens, Houten, The Netherlands), or a PalmSens4 (PalmSens, Houten, The Netherlands) combined with a MUX8-R2 multiplexer (PalmSens, Houten, The Netherlands) controlled by PSTrace/MultiTrace software. Unmodified Italsens IS-C SPE (PalmSens, Houten, The Netherlands) were used for all experiments. The IS-C SPE contains a carbon working electrode ($\varnothing$ = 3 mm), a carbon counter electrode, and a silver reference electrode. All potentials reported in this work are versus Ag pseudoreference electrode. Square wave voltammetric (SWV) measurements were performed in Britton–Robinson buffer with 0.1 M KCl by depositing 100 µL of the sample solution on the SPE. The samples were allowed to interact with the electrode surface for five minutes before the measurements were started. Instrumental parameters were: 60 mV amplitude; 25 Hz frequency; 5 mV potential step. All SWVs obtained were baseline corrected using a mathematical algorithm "moving average" (peak width = 1) in PSTrace software to improve the resolution of the peaks over the background.

2.3. Cocaine Detection in Oral Fluid

OF samples were collected from healthy volunteers from the research group immediately before analysis. Samples were collected at least 2 h after food consumption or taking any medication. OF collection was performed in two manners: (i) by expectoration in a 3 mL testing tube, or (ii) by using an Intercept i2 oral fluid collection device (OraSure Technologies). The OF samples were diluted in Britton–Robinson buffer solution (pH 10, unless specified otherwise) containing SDS at the desired dilution factor before electrochemical interrogation.

3. Results and Discussion

3.1. Analytical Characterization of Cocaine in Buffer Solution

In the first step, the electrochemical behavior of cocaine in buffer solution (pH 9) was studied. As our group had previously shown that the use of the surfactant SDS can enhance the cocaine signal, SDS was added to the buffer solution (0.075 mg mL^{-1}) [9]. Thus, a cocaine oxidation peak was observed around +0.85 V. The performance of the detection method was assessed by executing a calibration curve (I_p (µA) = 1.41 × $c_{cocaine}$ (µM) − 0.88) with concentrations in the range from 0.1 to 10 µM (Figure 2). While the limit of detection (LOD) of 1.0 µM is not adequate for roadside drug testing, it is similar to LODs reported for the direct electrochemical detection of cocaine using SWV by other authors [10,11].

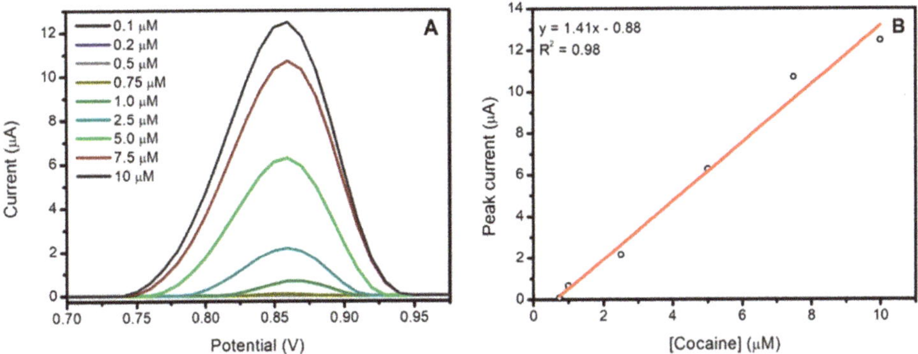

Figure 2. Analytical characterization of cocaine in buffer solution under optimal conditions. (**A**) SWVs of increasing cocaine concentrations from 0.1 µM to 10 µM, (**B**) corresponding calibration curve.

3.2. Study of the OF Matrix Effects

Before the electrochemical method was tested in OF, the influence of several compounds present in OF on the cocaine signal was investigated. Whole saliva is a complex heterogeneous mixture containing proteins, electrolytes and small organic compounds and is rich in antioxidants [12,13]. The effect of the anti-oxidants uric acid (200 μM) and ascorbate (vitamin C, 5 μM), as well as that of urea (4.5 mM) in a binary mixture with 5 μM cocaine in a buffer solution containing 0.075 mg mL^{-1} SDS was evaluated using SWAdSV. The concentrations of the potential interferents were selected according to the regular physiological levels found in OF [14]. The voltammograms showed that ascorbate and urea were not electroactive under the experimental conditions (Figure 3A). The voltammogram of uric acid showed an oxidation peak at +0.12 V, with a shoulder around +0.3 V. The presence of all three compounds resulted in a decrease in peak current for cocaine as compared to a 5 μM cocaine reference sample. This decrease was highest for urea, with a 21% loss in peak current.

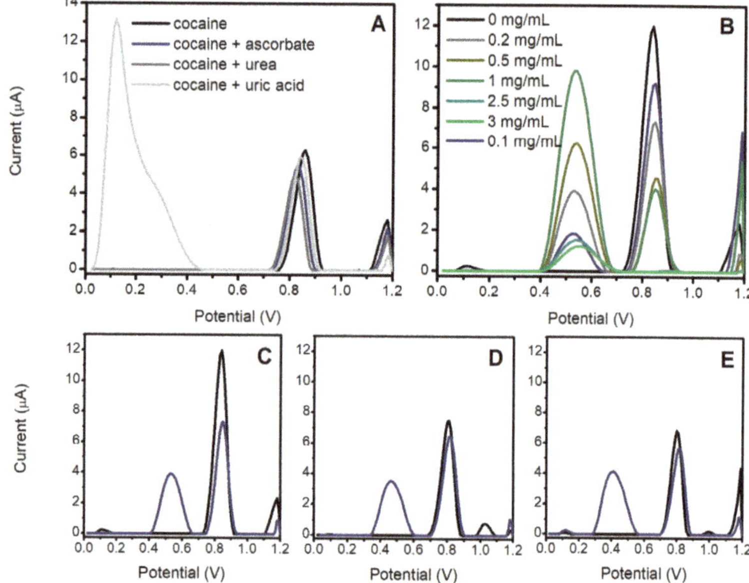

Figure 3. Investigation of the OF matrix effects on 10 μM cocaine under optimal conditions: (**A**) Effect of constituents: ascorbate, urea, uric acid and (**B**) effect of albumin concentration. Effect of the pH on the oxidation signal for 10 μM cocaine in the presence of 0.2 mg mL^{-1} albumin with (**C**) pH 9, (**D**) pH 10, and (**E**) pH 11. Black line: 10 μM cocaine, blue line: 10 μM cocaine in the presence of 0.2 mg mL^{-1} albumin. All tests were executed in BR buffer containing 0.075 mg mL^{-1} SDS.

Proteins are known to have biofouling effects on electrodes due to non-specific adsorption on the surface of the biochemical sensor [15]. This can result in the decrease in performance of the biosensor and loss in sensitivity and specificity for the target analyte. As saliva can contain over 1000 different peptides and proteins, the detection of cocaine in OF is expected to be hindered by biofouling effects [16]. To investigate this biofouling phenomenon, albumin was selected as a model protein because it is the most abundant protein in biofluids [17]. First, the effect of different albumin concentrations (0, 0.1, 0.2, 0.5, 1, 2.5, and 3 mg mL^{-1}) on the electrochemical signal for 10 μM cocaine was evaluated (Figure 3B). Albumin was shown to be electroactive, with an oxidation peak at +0.54 V. Interestingly, the albumin peak decreased at concentrations above 2.5 mg mL^{-1}. As expected, the cocaine peak current decreased with an increase in albumin concentration.

Unfortunately, at an albumin concentration of 3 mg mL^{-1}, the cocaine peak was completely suppressed. However, average levels of albumin in OF are ca. 0.9 mg mL^{-1} [18], which should still allow for the electrochemical detection of cocaine.

In a strategy to minimize the biofouling effects, the effect of pH was studied (Figure 3C–E). As the pKa values of the α-amino hydrogen of amino acids are in the range 8.72–10.70 [19], it was predicted that at pH 11 albumin would be negatively charged. Therefore, a repulsion by the negatively charged SDS moieties is expected. The SDS/SPE was tested with 10 μM cocaine and a binary mixture of 10 μM cocaine with 0.2 mg mL^{-1} albumin in buffer solutions of pH 9, pH 10, and pH 11. The albumin peak potential shifted towards less positive values with increased pH, as its deprotonated form is easier to oxidize. While the decrease in the electrochemical signal was 1.6-fold at pH 9, it was only 1.2-fold for pH 10 and pH 11. This indicated that less albumin was adsorbed at the electrode surface as predicted. It is important to note that the concentration of albumin and other proteins in OF might vary between individuals. To obtain a more reproducible method of cocaine detection, it is therefore important to minimize the biofouling effects of proteins. Hence, pH 10 was selected as a compromise situation between reduced biofouling effects of albumin compared to buffer pH 9, and a higher peak current compared to buffer pH 11.

3.3. Investigation of Two OF Collection Methods

Direct electrochemical measurement in OF is possible, as OF contains electrolytes and is ionically conductive [20]. However, since the OF composition and pH of individuals may vary, it is preferable to add a buffer solution in order to control the chemical composition and pH of the solution. Dilution of the buffer solution has the disadvantage that the concentration of the illicit drug in the total solution decreases, but it can cope with strong interferences. A dilution factor of 1:5 (OF:buffer) was selected as a compromise between the decrease in cocaine concentration, decrease in background effects, and stability of sample pH.

As OF collection by expectoration is slow and unpleasant for donors, and also suffers from a lack of hygiene, OF collection by a commercial device was explored. The Intercept i2 was used as a model device. The Intercept i2 works by placing it under the tongue of the donor until the indicator turns blue and the desired amount of is collected. According to the manufacturer, this device collects 1 mL of in an average time of 3–4 min [21]. To test the performance of the devices, several experiments were carried out to determine (i) the time of OF collection, (ii) the amount of OF collected, (iii) the amount of OF extracted, and (iv) the recovery of cocaine. First, the amount of collected and the collection time were examined. The OF from two individuals was collected three times using the Intercept i2. Before and immediately after collection, the devices were weighed. On average, 1.2 g of OF was collected. Assuming a density of 1 g mL^{-1}, this amounts to approximately 1.2 mL of OF, which is more than the manufacturer claimed. The average collection time was just over 1 min, which is substantially shorter than the waiting time mentioned by the manufacturer.

Three different approaches were explored for the recovery of cocaine and extraction of OF from the Intercept devices. OF was collected by expectoration and spiked with 10 μM cocaine. A small amount of spiked OF was put aside for comparison, while the remaining sample was collected with the Intercept device. The recovery of cocaine was examined by comparing the voltammograms obtained from the recovered samples with a reference voltammogram from the OF that was set aside (Figure 4A–C). In these experiments, the recovered fluids were diluted with 2 mL of buffer solution containing 0.075 mg mL^{-1} SDS (dilution factor 1:2). It was assumed that 1 mL of OF was collected with the Intercept devices, as this is the amount stated by the manufacturer and what laboratories work within their analyses. In the first cocaine recovery approach, the preservation liquid in the Intercept i2 collection vial was removed and replaced with the buffer solution. After the OF was collected with the Intercept device, the collection pad was placed in the collection vial. The vial was vigorously shaken, and then left to rest for 5 min so that diffusion could take place, before the liquids in the collection vial were collected and analyzed with SWV. After recovery from the Intercept device, the peak currents for cocaine and albumin decreased 2.4-fold and 2.1-fold, respectively (Figure 4A). This indicated that the OF and cocaine recovery from the device was not complete. The second approach of recovery consisted of centrifugation at 3000 rpm for 5 min, as this is recommended in the manual by the manufacturer. After centrifugation, buffer solution was added to the recovered fluids. The voltammogram of the recovered cocaine showed a more intense peak than the reference voltammogram (Figure 4B). This could have been due to evaporation of the sample during the manipulation leading to an error in comparison to spiked OF. Recovery by centrifugation has the disadvantage that it makes the total procedure for roadside testing more difficult and more expensive. Therefore, as an alternative approach, the recovery was performed by pressing the collection pad using a syringe. To do this, the Intercept i2 collection device was broken open and the collection pad was removed. The recovered fluids were collected in a tube and mixed with the buffer solution. The voltammograms showed a slight decrease (9%) in peak current for cocaine (Figure 4C). The peak current for albumin decreased with 26%, indicating that albumin might be more retained at the collection pad than cocaine. As the change in peak current for cocaine was smallest when the recovery was performed using a syringe, this strategy was chosen as an optimal procedure for cocaine recovery. The recovery of cocaine in this approach was tested using three Intercept devices (Figure 4D–F). On average, 57.8 ± 4.8%) of the collected OF was recovered. For all three collection devices, the peak current of the recovered cocaine was higher than the peak current for cocaine in the reference sample. The increase in peak current was largest for device 1 (1.5-fold increase), which was also the device from which most oral fluid was recovered.

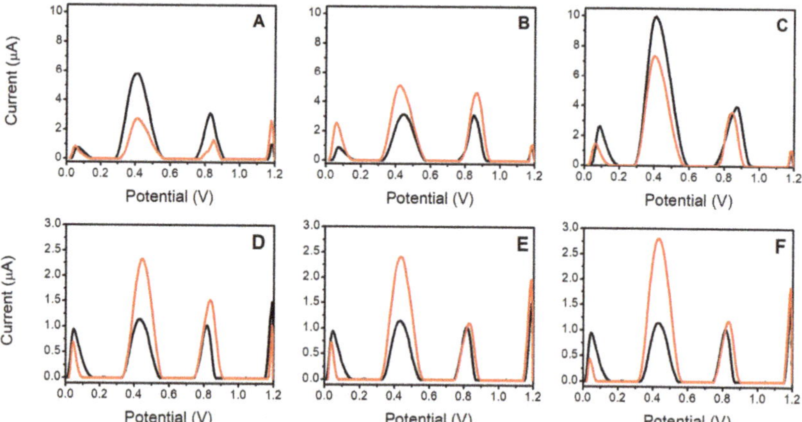

Figure 4. Recovery of 10 μM cocaine from the Intercept i2 collection device by (**A**) vigorously shaking, (**B**) centrifugation for 5 min at 3000 rpm, and (**C**) removing the collection pad from the device and pressing it using a syringe. Recovery of 10 μM cocaine from three Intercept i2 devices using the syringe method for (**D**) device 1, (**E**) device 2, and (**F**) device 3. Black line: 10 μM cocaine in OF for reference, red line: recovery of 10 μM cocaine from the Intercept i2 device. All SWVs for the recovery study were performed with 3-fold diluted OF in Britton–Robinson buffer pH 10 containing 0.075 mg mL^{-1}.

4. Conclusions

In this work, a rapid, inexpensive, and sensitive electrochemical method for the detection of cocaine in OF was explored. In buffer solution, the LOD for cocaine detection was found to be 1 μM. For the first time, the interference of endogenous compounds present in the OF matrix on the electrochemical detection of cocaine was studied. Albumin showed to have fouling effects on the electrode, causing a decrease in the sensitivity. The antifouling effects were successfully reduced by adjusting the pH of the buffer solution from pH 9 to 10. A sampling method for the direct measurement in OF was developed and integrated with the SDS/SPE system, as a first step towards the application of electrochemical methods for illicit drugs detection in OF in the field.

Author Contributions: Conceptualization, F.J. and M.P.; methodology, F.J.; validation, F.J.; formal analysis, F.J.; investigation, F.J. and M.P.; resources, K.D.W.; data curation, F.J. and M.P.; writing—original draft preparation, F.J.; writing—review and editing, F.J., M.P. and K.D.W.; visualization, F.J. and M.P.; supervision, M.P. and K.D.W.; project administration, K.D.W.; funding acquisition, K.D.W. All authors have read and agreed to the published version of the manuscript.

Funding: This research was funded by the FWO NRF Bilateral Scientific Cooperation South Africa (grant number G0F9820N) in the project Electrochemistry and nanostructured electrocatalysts for tackling substance abuse. The authors also acknowledge financial support from the University of Antwerp (IOF).

Institutional Review Board Statement: Not applicable.

Informed Consent Statement: Not applicable.

Data Availability Statement: Not applicable.

Conflicts of Interest: The authors declare no conflict of interest.

References

1. *World Drug Report 2021*; United Nations Office on Drugs and Crime (UNODC): Vienna, Austria, 2021.
2. *World Drug Report 2020*; United Nations Office on Drugs and Crime (UNODC): Vienna, Austria, 2020.
3. *European Drug Report 2020: Trends and Developments*; European Monitoring Centre for Drugs and Drug Addiction (EMCDDA): Lisbon, Portugal, 2020.
4. *Drug Use and Road Safety*; World Health Organization (WHO): Geneva, Switzerland, 2016.
5. Schulze, H.; Schumacher, M.; Urmeew, R.; Auerbach, K.; Alvarez, J.; Bernhoft, I.M.; de Gier, H.; Hagenzieker, M.; Houwing, S.; Knoche, A.; et al. *Driving Under the Influence of Drugs, Alcohol and Medicines in Europe—Findings from the DRUID Project*; Publications Office of the European Union: Luxembourg, 2012.
6. Scientific Working Group for the Analysis of Seized Drugs (SWGDRUG). Scientific Working Group for the Analysis of Seized Drugs (SWGDRUG) Recommendations. Available online: https://www.swgdrug.org/Documents/SWGDRUG%20Recommendations%20Version%208_FINAL_ForPosting_092919.pdf (accessed on 24 June 2021).
7. Ahmed, S.R.; Chand, R.; Kumar, S.; Mittal, N.; Srinivasan, S.; Rajabzadeh, A.R. Recent Biosensing Advances in the Rapid Detection of Illicit Drugs. *TrAC Trends Anal. Chem.* **2020**, *131*, 116006. [CrossRef]
8. Posthuma-Trumpie, G.A.; Korf, J.; Van Amerongen, A. Lateral Flow (Immuno)Assay: Its Strengths, Weaknesses, Opportunities and Threats. A Literature Survey. *Anal. Bioanal. Chem.* **2009**, *393*, 569–582. [CrossRef] [PubMed]
9. Parrilla, M.; Joosten, F.; De Wael, K. Enhanced Electrochemical Detection of Illicit Drugs in Oral Fluid by the Use of Surfactant-Mediated Solution. *Sens. Actuators B Chem.* **2021**, *348*, 130659. [CrossRef]
10. De Jong, M.; Sleegers, N.; Kim, J.; Van Durme, F.; Samyn, N.; Wang, J.; De Wael, K. Electrochemical Fingerprint of Street Samples for Fast On-Site Screening of Cocaine in Seized Drug Powders. *Chem. Sci.* **2016**, *7*, 2364–2370. [CrossRef] [PubMed]
11. Rocha, R.G.; Stefano, J.S.; Arantes, I.V.S.; Ribeiro, M.M.A.C.; Santana, M.H.P.; Richter, E.M.; Munoz, R.A.A. Simple Strategy for Selective Determination of Levamisole in Seized Cocaine and Pharmaceutical Samples Using Disposable Screen-Printed Electrodes. *Electroanalysis* **2019**, *31*, 153–159. [CrossRef]
12. Del Vigna de Almeida, P.; Trindade Grégio, A.M.; Naval Machado, M.Â.; Adilson Soares de Lima, A.; Reis Azevedo, L. Saliva Composition and Functions: A Comprehensive Review. *J. Contemp. Dent. Pract.* **2008**, *9*, 72–80.
13. Battino, M.; Ferreiro, M.S.; Gallardo, I.; Newman, H.N.; Bullon, P. The Antioxidant Capacity of Saliva. *J. Clin. Periodontol.* **2002**, *29*, 189–194. [CrossRef] [PubMed]
14. Rehak, N.N.; Cecco, S.A.; Csako, G. Biochemical Composition and Electrolyte Balance of "unstimulated" Whole Human Saliva. *Clin. Chem. Lab. Med.* **2000**, *38*, 335–343. [CrossRef] [PubMed]
15. Russo, M.J.; Han, M.; Desroches, P.E.; Manasa, C.S.; Dennaoui, J.; Quigley, A.F.; Kapsa, R.M.I.; Moulton, S.E.; Guijt, R.M.; Greene, G.W.; et al. Antifouling Strategies for Electrochemical Biosensing: Mechanisms and Performance toward Point of Care Based Diagnostic Applications. *ACS Sens.* **2021**, *6*, 1482–1507. [CrossRef] [PubMed]
16. Inzitari, R.; Cabras, T.; Rossetti, D.V.; Fanali, C.; Vitali, A.; Pellegrini, M.; Paludetti, G.; Manni, A.; Giardina, B.; Messana, I.; et al. Detection in Human Saliva of Different Statherin and P-B Fragments and Derivatives. *Proteomics* **2006**, *6*, 6370–6379. [CrossRef] [PubMed]

17. Deutsch, O.; Fleissig, Y.; Zaks, B.; Krief, G.; Aframian, D.J.; Palmon, A. An Approach to Remove Alpha Amylase for Proteomic Analysis of Low Abundance Biomarkers in Human Saliva. *Electrophoresis* **2008**, *29*, 4150–4157. [CrossRef] [PubMed]
18. Shaila, M.; Pai, G.P.; Shetty, P. Salivary Protein Concentration, Flow Rate, Buffer Capacity and PH Estimation: A Comparative Study among Young and Elderly Subjects, Both Normal and with Gingivitis and Periodontitis. *J. Indian Soc. Periodontol.* **2013**, *17*, 42–46. [CrossRef] [PubMed]
19. Vanderbilt University. Amino Acids. Available online: https://www.vanderbilt.edu/AnS/Chemistry/Rizzo/stuff/AA/AminoAcids.html (accessed on 25 June 2021).
20. Aframian, D.; Davidowitz, T.; Benoliel, R. The Distribution of Oral Mucosal PH Values in Healthy Saliva Secretors. *Oral Dis.* **2006**, *12*, 420–423. [CrossRef] [PubMed]
21. OraSure Technologies. Oral Fluid Drug Testing. Available online: https://www.orasure.com/products-substance-abuse/i2.html (accessed on 25 June 2021).

engineering proceedings

Abstract

Development of Electrochemical Sensors Based on Electrosynthesized Imprinted Polymers for Cobalt (Co^{2+}) Ion Determination in Water [†]

Nelson Arturo Manrique-Rodriguez, Sabrina Di Masi * and Cosimino Malitesta

Laboratorio di Chimica Analitica, Dipartimento di Scienze e Tecnologie Biologiche ed Ambientali, Università del Salento, Via Monteroni, 73100 Lecce, Italy; nelsonarturo.manriquerodriguez@studenti.unisalento.it (N.A.M.-R.); cosimino.malitesta@unisalento.it (C.M.)

* Correspondence: sabrina.dimasi@unisalento.it

† Presented at the 2nd International Electronic Conference on Biosensors, 14–18 February 2022; Available Online: https://sciforum.net/event/IECB2022.

Abstract: Preliminary results on an electrosynthesized ion-imprinted polymeric (IIP) film for the development of a Co^{2+} sensor are reported herein. The sensor was prepared by CV electropolymerization of 2-aminophenol (2-AP) monomer in the presence of Co^{2+} ions, which acted as the template. The screen-printed carbon electrodes (SPCEs) were used as transducers during sensor development, whereas the cyclic voltammetry (CV) and electrochemical impedance spectroscopy (EIS) were used for the electrochemical characterization of sensors and for Co^{2+} ion sensing, respectively. The CV (potential range −0.2 and 1.2 V) and EIS measurements were performed in PBS (pH 7.8 , 0.1 M) containing 0.1 mol L^{-1} KCl solution and 5.0 mmol L^{-1} of $Fe(CN)_6^{3-}/^{4-}$ as the redox probe ; for EIS an open circuit and data were settled through a sinusoidal potential perturbation of 0.01 V amplitude and 57 as frequency values that were logarithmically distributed over a range of frequencies between 0.01 Hz and 100 kHz. A not imprinted polymer (NIP) was prepared as a control under the same protocol, but without adding the template into the polymerization mixture. In these preliminary tests, the electropolymerization patterns of IIP polymers were found to be consistent with the findings previously reported. After electropolymerization, rinsed electrodes were incubated in different Co^{2+} concentrations of ions to be tested through EIS showing a response in the range 1–8 μM. A multivariate optimization based on the design of experiment (DOE) was employed to study the effect of parameters on electrochemical performances of the sensor.

Keywords: ion-imprinted polymer; 2-AP; electrochemical sensor; Co^{2+} ions; electropolymerization

Citation: Manrique-Rodriguez, N.A.; Di Masi, S.; Malitesta, C. Development of Electrochemical Sensors Based on Electrosynthesized Imprinted Polymers for Cobalt (Co^{2+}) Ion Determination in Water. *Eng. Proc.* **2022**, *16*, 15. https://doi.org/10.3390/IECB2022-12281

Academic Editors: Giovanna Marrazza and Sara Tombelli

Published: 15 February 2022

Publisher's Note: MDPI stays neutral with regard to jurisdictional claims in published maps and institutional affiliations.

Copyright: © 2022 by the authors. Licensee MDPI, Basel, Switzerland. This article is an open access article distributed under the terms and conditions of the Creative Commons Attribution (CC BY) license (https://creativecommons.org/licenses/by/4.0/).

Supplementary Materials: The following supporting information can be downloaded at: https://www.mdpi.com/article/10.3390/IECB2022-12281/s1.

Author Contributions: Conceptualization, S.D.M. and C.M.; methodology, S.D.M. and C.M.; formal analysis, N.A.M.-R.; investigation, N.A.M.-R.; data curation, S.D.M., C.M. and N.A.M.-R.; writing—original draft preparation, S.D.M. and N.A.M.-R.; writing—review and editing, S.D.M. and C.M.; supervision, S.D.M. and C.M. All authors have read and agreed to the published version of the manuscript.

Funding: This research was funded by the project "CASCADE" (014-2020 Interreg V-A IT-HR CBC "strategic" project ID 10255941).

Institutional Review Board Statement: Not applicable.

Conflicts of Interest: The authors declare no conflict of interest.

Proceeding Paper

Fast and Accurate Determination of Minute Ochratoxin A Levels in Cereal Flours: Towards Application at the Field [†]

Chrysoula-Evangelia Karachaliou [1,*], Georgios Koukouvinos [2], Katerina Pissaridi [3], Dimitris Ladikos [3], Dimitris Goustouridis [4], Ioannis Raptis [4], Evangelia Livaniou [1], Sotirios Kakabakos [2] and Panagiota Petrou [2]

[1] Immunopeptide Chemistry Lab, INRASTES, National Centre for Scientific Research "Demokritos", P.O. Box 60037, 15310 Agia Paraskevi, Greece; livanlts@rrp.demokritos.gr
[2] Immunoassay/Immunosensors Lab, INRASTES, National Centre for Scientific Research "Demokritos", P.O. Box 60037, 15310 Agia Paraskevi, Greece; geokoukoubinos@yahoo.gr (G.K.); skakab@rrp.demokritos.gr (S.K.); ypetrou@rrp.demokritos.gr (P.P.)
[3] Yiotis S.A., 130 Kifisou Ave., 12131 Peristeri, Greece; kpissaridi@jotis.gr (K.P.); dladikos@jotis.gr (D.L.)
[4] ThetaMetrisis S.A., 14 Polydefkous, 12243 Egaleo, Greece; dgoustouridis@thetametrisis.com (D.G.); raptis@thetametrisis.com (I.R.)
* Correspondence: xrisak15@hotmail.com
[†] Presented at the 2nd International Electronic Conference on Biosensors, 14–18 February 2022; Available online: https://sciforum.net/event/IECB2022.

check for **updates**

Citation: Karachaliou, C.-E.; Koukouvinos, G.; Pissaridi, K.; Ladikos, D.; Goustouridis, D.; Raptis, I.; Livaniou, E.; Kakabakos, S.; Petrou, P. Fast and Accurate Determination of Minute Ochratoxin A Levels in Cereal Flours: Towards Application at the Field. *Eng. Proc.* **2022**, *16*, 14. https://doi.org/10.3390/IECB2022-12270

Academic Editors: Giovanna Marrazza and Sara Tombelli

Published: 14 February 2022

Publisher's Note: MDPI stays neutral with regard to jurisdictional claims in published maps and institutional affiliations.

Copyright: © 2022 by the authors. Licensee MDPI, Basel, Switzerland. This article is an open access article distributed under the terms and conditions of the Creative Commons Attribution (CC BY) license (https://creativecommons.org/licenses/by/4.0/).

Abstract: Ochratoxins are a group of mycotoxins produced as secondary metabolites by several fungi of *Aspergillus* and *Penicillium* species. Ochratoxin A (OTA) is the most toxic member of the group and can be found in a large variety of widely consumed foods, such as coffee, cocoa, wine, and flour. Reliable determination of OTA levels in food samples is therefore indispensable to ensure compliance with MRLs set by national/European regulations and minimize health risks for consumers. In the current study, a label-free biosensor based on white light reflectance spectroscopy (WLRS) for the rapid and accurate determination of OTA in cereal flour samples is demonstrated. A Si chip with a 1-μm-thick thermal SiO_2 on top plays the role of transducer after the immobilization of an OTA–protein conjugate on its surface. For the assay, a mixture of an in-house-developed anti-OTA antibody with the calibrators or the samples is injected over the chip surface, followed by reaction with a secondary biotinylated antibody and streptavidin for signal amplification. The label-free, real-time monitoring of immunoreactions occurring on the SiO_2/Si chip surface is achieved by recording the shift in the reflected interference spectrum caused by the immunoreactions. This shift is converted through appropriate mathematical processing to an effective biomolecular adlayer thickness. After optimization, the sensor is capable of detecting OTA in wheat flour samples at concentrations as low as 60 pg/mL within 25 min. The assay is repeatable, with intra- and inter-assay CVs ≤5.9% and ≤9.0%, respectively. The assay's excellent analytical characteristics and short analysis time, in combination with the small size of the device, render the proposed WLRS system ideal for the quantitative determination of minute OTA levels at the point-of-need.

Keywords: mycotoxins; Ochratoxin A; flour; white light reflectance spectroscopy; label-free immunosensor; point-of-need

1. Introduction

Ochratoxin A (OTA) is a low-molecular-weight mycotoxin (403.81 Da; Scheme 1) produced as a secondary metabolite mainly by filamentous fungi of the genera *Aspergillus* and *Penicillium* [1]. Due to the colonization of these species in a plethora of food crops during cultivation, harvest and post-harvest procedures, OTA has been reported to be present in a series of highly consumed agricultural products, such as cereals, coffee, cocoa, beer and wine [2]. OTA is considered harmful for humans and animals, since there is evidence associating its consumption with chronic toxicity (genotoxicity, immunotoxicity, nephrotoxicity, hepatotoxicity, etc.), teratogenicity, mutagenicity and carcinogenicity. In accordance with this, the International Agency for Research on Cancer has classified OTA as group 2B—a possible carcinogen to humans [3]. In order to control and minimize the public health risk, the European Union (EU) has established a strict maximum limit of 2 ng/mL for OTA in cereals and cereal flours, which was adopted by several non-EU countries as well [4].

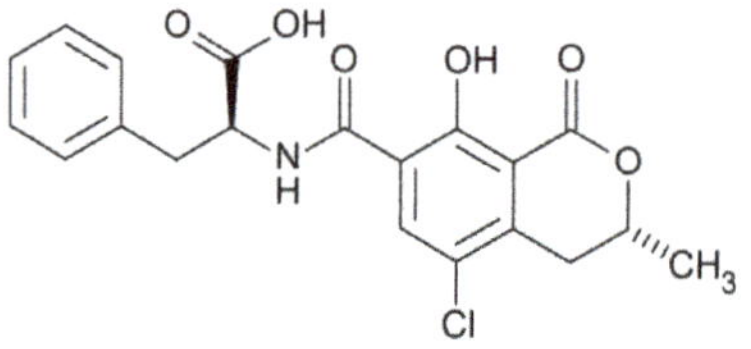

Scheme 1. Chemical structure of Ochratoxin A.

Nowadays, the analysis of OTA in foodstuffs, including flour, is performed by well-established analytical techniques, mainly high-performance liquid chromatography (HPLC) coupled to fluorescence or mass spectrometry detection [5,6]. These sophisticated techniques are characterized by high reliability, selectivity and sensitivity. However, the high analysis cost, the need for skilled personnel and the bulky instruments are major bottlenecks for the application of these techniques to the routine high-throughput screening and/or point-of-need (PoN) analysis of OTA. In this direction, in the last decade(s), biosensors have raised great expectations as an emerging technology with potential for automation, miniaturization and incorporation into portable setups for the rapid and reliable determination of OTA levels in flours across the production line and up to the retail shelf [7].

In the current work, we present the development of a sensitive and reliable method for the rapid immunochemical detection of OTA in cereal flours based on a white light reflectance spectroscopy (WLRS) optical sensor. The WLRS methodology involves a visible/near-infrared light source, a miniaturized USB controlled spectrometer operating in the 400–750 nm spectral range and a reflection probe of seven optical fibers, of which six are arranged at the periphery of the probe and one at its center. The white light emitted from the light source is guided through the six fibers at the periphery of the reflection probe to the bio-modified surface, and at the same time, the central seventh fiber collects the reflected light and directs it to the spectrometer. The immunochemical reactions take place on top of a 1-μm-thick SiO_2 layer grown by thermal oxidation of silicon chips. Thus, the light reflected from the silicon surface passes through the SiO_2 and the biomolecular layer and an interference spectrum is received that shifts to higher wavelengths during the course of immunoreactions. The software calculates the effective thickness of the biomolecular layer that caused the spectral shift, and this thickness in nm is the sensor signal. For the determination of OTA, an indirect competitive immunoassay format realized in three steps was implemented. At first, mixtures of an in-house-developed rabbit anti-OTA antibody with OTA calibrators or samples were passed over an amino-silanized chip biofunctionalized with an OTA–protein conjugate; then, a biotinylated secondary anti-rabbit IgG antibody and streptavidin were applied to enhance the signal received by the primary immunoreaction. All assay parameters were optimized in terms of absolute signal, detection sensitivity and total analysis time, aiming at a fast method for the sensitive and reliable determination of OTA in cereal flours.

2. Materials and Methods

2.1. Reagents and Instrumentation

Ochratoxin A (OTA) and the OTA conjugate with ovalbumin (OTA-OVA) were purchased from Aokin AG (Berlin, Germany). Rabbit anti-OTA antibody (primary antibody) was in-house-developed. Biotinylated goat anti-rabbit IgG antibody (secondary antibody), streptavidin and (3-aminopropyl) triethoxysilane (APTES) were obtained from Sigma-

Aldrich (Darmstadt, Germany). Bovine serum albumin (BSA) was from Acros Organics (Geel, Belgium). IgG elution buffer was from Thermo Fisher Scientific Inc. (Waltham, MA, USA). All other chemicals were from Merck KGaA (Darmstadt, Germany). The RIDASCREEN® Ochratoxin A 30/15 enzyme immunoassay kit was purchased from R-Biopharm AG (Darmstadt, Germany). Flour samples were provided by Yiotis SA industry.

OTA calibrators with concentrations ranging from 0.05 to 200 ng/mL were prepared from a 2 mg/mL OTA stock solution in absolute ethanol after proper dilution in a 1:9 mixture of ethanol with 10 mM phosphate buffer, pH 7.4, 0.9 % (w/v) NaCl, 0.02 % (w/v) KCl, 0.2% (w/v) BSA (assay buffer).

Four-inch Si wafers were purchased from Si-Mat Germany (Kaufering, Germany). The visible/near-infrared light source of the sensor was a product of ThetaMetrisis SA (Athens, Greece); the miniaturized USB-controlled spectrometer and the reflection probe were obtained from Ocean Insight (Duiven, The Netherlands).

2.2. Chip Biofunctionalization and Assay Protocol

For chip biofunctionalization, an OTA conjugate with OVA was deposited on APTES-modified chips [8] and incubated overnight at RT. The next day, after proper washing with phosphate buffer (washing buffer), the chips were blocked in 2% (w/v) BSA solution in washing buffer for 3 h, washed once more with washing buffer and distilled water, dried under nitrogen flow and used for the assay.

The fluidic module was applied to each biofunctionalized chip, and assay buffer was run to acquire a stable baseline. The assay was performed by flowing 1:1 volume mixtures of calibrators (0.05–200 ng/mL in assay buffer) or flour extracts 2-times diluted with assay buffer, with the rabbit anti-OTA antibody (1 µg/mL in assay buffer) for 15 min (primary immunoreaction). Next, a biotinylated anti-rabbit IgG solution (secondary immunoreaction, 7 min) and a streptavidin solution (3 min) were introduced. Lastly, IgG elution buffer was run over the chip to remove all molecules bound to the OTA–OVA conjugate, followed by assay buffer for surface re-equilibration. The reagent's flow rate throughout the experiment was 50 µL/min. To prepare the calibration curve, the effective thickness of the biomolecular layer (signal) determined for the calibrators (S_x) was expressed as a percentage of the zero-calibrator signal-maximum signal (S_0) and plotted against the analyte concentration in the calibrators.

3. Results and Discussion

3.1. WLRS Assay Optimization

Many assay parameters had to be optimized to establish the sensitive and rapid determination of OTA with the WLRS sensor. Due to the competitive nature of the immunoassay, at first, titration experiments were carried out employing different concentrations of OTA conjugate (50–500 µg/mL) combined with different concentrations of the anti-OTA antibody (0.5–4 µg/mL). The combination that provided a satisfactory analytical signal and assay sensitivity was 200 µg/mL for the OTA conjugate and 1 µg/mL for the anti-OTA antibody. Another parameter to optimize was the duration of the different assay steps with the condition that the total analysis time should be as short as possible. The assay duration was 25 min; it was 15 min for the primary immunoreaction, 7 min for the secondary one and 3 min for the biotinylated secondary antibody and streptavidin reaction step (Figure 1).

3.2. WLRS Assay Characteristics

A representative OTA calibration curve obtained with the WLRS immunosensor and calibrators prepared in assay buffer following the final assay protocol is depicted in Figure 2. It should be noted that an identical calibration curve was obtained when the extract from a wheat flour that did not contain detectable concentrations of OTA diluted 1:1 v/v with assay buffer was used as a matrix for the preparation of OTA calibrators.

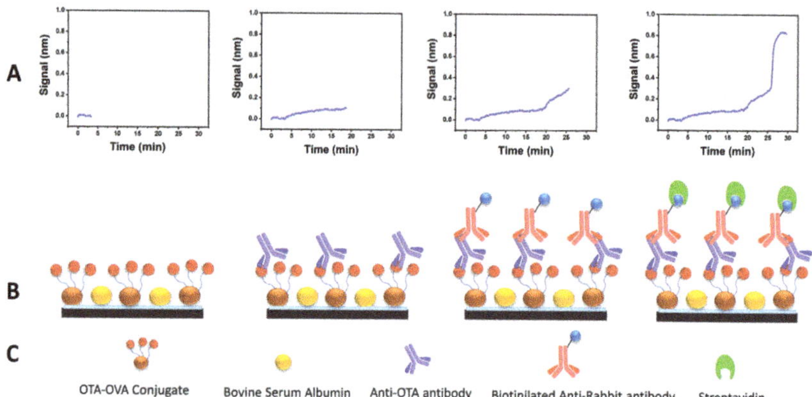

Figure 1. (**A**) Real-time sensor response at each OTA WLRS immunoassay step; (**B**) schematic representation of the main immunoassay steps; (**C**) main immunoassay reagents.

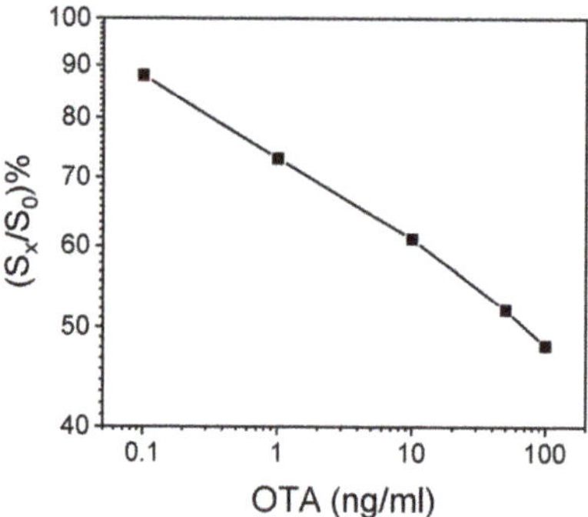

Figure 2. Typical calibration curve obtained with OTA calibrators prepared in assay buffer. Each point represents the mean value of three independent runs $\pm$ SD. S_0 = zero-calibrator signal; S_x = calibrator signal.

The proposed immunosensor was sensitive, with an assay limit of detection (LoD) of 30 pg/mL, which corresponded to the mean signal value -3SD of 15 replicates of the zero calibrator. The assay dynamic range extended up to 200 ng/mL. In addition, the assay was reproducible, with intra-assay co-efficient of variation (CV) values $\leq$5.9%, determined by running three control samples in triplicate within the same day, and inter-assay CVs $\leq$9.0%, determined by duplicate measurements of the three controls on seven different days.

The OTA assay developed has been applied to the determination of OTA levels in a small number of cereal flour samples. There was a very good correlation of the OTA concentrations determined in these samples with the WLRS assay with the OTA values determined in the same samples using a commercially available immunoassay kit. Evaluation with a larger number of cereal flour samples is underway.

4. Conclusions

In the current work, a WLRS-based biosensing platform was applied to the label-free and real-time immunochemical determination of OTA. The developed sensor enabled the fast (25 min) and sensitive quantification of OTA levels at concentrations as low as 30 pg/mL. The combination of a series of assets, i.e., the high analytical sensitivity, the short analysis time and the small instrument size, resulted in a bioanalytical platform ideal for the quantitative determination of minute OTA levels at the point-of-need.

Author Contributions: Conceptualization, C.-E.K. and G.K.; methodology, I.R., E.L., S.K. and P.P.; software, D.G.; validation, C.-E.K. and G.K.; formal analysis, C.-E.K. and G.K.; investigation, C.-E.K. and G.K.; resources, K.P. and D.L.; data curation, I.R., E.L., S.K. and P.P.; writing—original draft preparation, C.-E.K. and G.K.; writing—review and editing, I.R., E.L., S.K. and P.P.; visualization, I.R.; supervision, P.P. All authors have read and agreed to the published version of the manuscript.

Funding: The work was partially funded by the EU Horizon 2020 project "DIgital TEChnologies as an enabler for a conTinuous transformation of food safety system (DITECT)", contract number 861915.

Institutional Review Board Statement: Not applicable.

Informed Consent Statement: Not applicable.

Data Availability Statement: The data presented in this study are available on request from the corresponding author. The data are not publicly available due to privacy issues.

Conflicts of Interest: The authors declare no conflict of interest.

References

1. Weidenbörner, M. *Encyclopedia of Food Mycotoxins*, 1st ed.; Springer: Berlin/Heidelberg, Germany, 2001.
2. Jørgensen, K. Occurrence of ochratoxin A in commodities and processed food—A review of EU occurrence data. *Food Addit. Contam.* **2005**, *22* (Suppl. 1), 26–30. [CrossRef] [PubMed]
3. Ostry, V.; Malir, F.; Toman, J.; Grosse, Y. Mycotoxins as human carcinogens-the IARC Monographs classification. *Mycotoxin Res.* **2017**, *33*, 65–73. [CrossRef] [PubMed]
4. Commission Regulation (EC) No. 1881/2006 setting maximum levels for certain contaminants in foodstuffs. *Off. J. Eur. Union* **2006**, *L 364/5*, 5–24.
5. Skarkova, J.; Ostry, V.; Malir, F.; Roubal, T. Determination of Ochratoxin A in Food by High Performance Liquid Chromatography. *Anal. Lett.* **2013**, *46*, 1495–1504. [CrossRef]
6. Li, P.; Zhang, Z.; Hu, X.; Zhang, Q. Advanced hyphenated chromatographic-mass spectrometry in mycotoxin determination: Current status and prospects. *Mass Spectrom. Rev.* **2013**, *32*, 420–452. [CrossRef] [PubMed]
7. Srinivasan, B.; Tung, S. Development and Applications of Portable Biosensors. *J. Lab. Autom.* **2015**, *20*, 365–389. [CrossRef] [PubMed]
8. Tsougeni, K.; Ellinas, K.; Koukouvinos, G.; Petrou, P.S.; Tserepi, A.; Kakabakos, S.E.; Gogolides, E. Three-dimensional (3D) plasma micro-nanotextured slides for high performance biomolecule microarrays: Comparison with epoxy-silane coated glass slides. *Colloids Surf. B Biointerfaces* **2018**, *165*, 270–277. [CrossRef] [PubMed]

engineering proceedings

MDPI

Proceeding Paper

Computational and Experimental Investigation of Microfluidic Chamber Designs for DNA Biosensors [†]

Sotiria D. Psoma [1,*], Ihor Sobianin [1] and Antonios Tourlidakis [2]

1 School of Engineering & Innovation, The Open University, Milton Keynes MK7 6AA, UK; ihor.sobianin@open.ac.uk
2 School of Engineering, University of Western Macedonia, 50100 Kozani, Greece; atourlidakis@uowm.gr
* Correspondence: sotiria.psoma@open.ac.uk; Tel.: +44-(0)1908-652629
† Presented at the 2nd International Electronic Conference on Biosensors, 14–18 February 2022; Available online: https://sciforum.net/event/IECB2022.

Abstract: A critical characteristic for continuous monitoring using DNA biosensors is the design of the microfluidics system used for sample manipulation, effective and rapid reaction and an ultra-low detection limit of the analyte. The selection of the appropriate geometrical design and control of microfluidic parameters are highly important for the optimum performance. In the present study, a number of different shapes of microchambers are designed and computationally assessed using a Multiphysics software. Flow parameters such as pressure drop, and shear rates are compared. Three-dimensional printing was used to construct the designs and an experimental investigation is underway for the validation of the computational results.

Keywords: microchamber design; microfluidics; simulation; DNA biosensor; 3D printing

Citation: Psoma, S.D.; Sobianin, I.; Tourlidakis, A. Computational and Experimental Investigation of Microfluidic Chamber Designs for DNA Biosensors. *Eng. Proc.* **2022**, *16*, 16. https://doi.org/10.3390/IECB2022-12252

Academic Editors: Giovanna Marrazza and Sara Tombelli

Published: 14 February 2022

Publisher's Note: MDPI stays neutral with regard to jurisdictional claims in published maps and institutional affiliations.

Copyright: © 2022 by the authors. Licensee MDPI, Basel, Switzerland. This article is an open access article distributed under the terms and conditions of the Creative Commons Attribution (CC BY) license (https://creativecommons.org/licenses/by/4.0/).

1. Introduction

Continuous monitoring is actively used in the medical practice as it allows for the constant monitoring of biomarkers. This facilitates timely and accurate diagnosis, which is followed by medical treatment. Genome sequencing is a method that deciphers huge amounts of data that are present in a DNA samples and is vital for detecting mutations and pathogens [1]. The analyte can be studied with the utilisation of a biosensor, a device which converts biochemical reaction into readable signal [2].

There are many methods of sample manipulation [3] used in laboratories, one of which is microfluidics. It has gained significant interest over recent years and has become an integral part of modern biosensors [4]. Among the many advantages of microfluidics are rapid reaction and ultra-low detection limit of the analyte [5].

In order to conduct the analysis, the microfluidic structure should provide a uniform fluid flow. One way of achieving this is changing the internal geometry of a microfluidic chamber. In this study, several computational simulations in COMSOL Multiphysics were conducted to assess how different chamber shapes influence the flow parameters. Based on the results, a number of recommendations regarding the microfluidics chamber design are proposed.

2. Materials and Methods

In the present work, nine different chamber designs were simulated and compared quantitatively as shown in Figure 1.

The physical domain that was analysed for every chamber design consisted of a typical microfluidic domain with an inlet section, a chamber and an outlet section as it is depicted in Figure 2. In the simulations, the fluid flow had an initial velocity of 0.01 m/s and was considered to be laminar. The fluid used in this scenario had a density of 1080 kg/m^3 and a dynamic viscosity of 1.75 mPa·s. The total length of all microfluidics configurations that were analysed was constant and equal to 26 mm, except for the focus flow chamber design which had a length of 100 mm. The thickness of all domains was equal to 2 mm and sharp edges were smoothened using fillets with radii of 1 mm. Seven cases were analysed for determining the influence of the chamber design, one case featured the trifurcation of chambers and one case included multiplication of inlets and outlets.

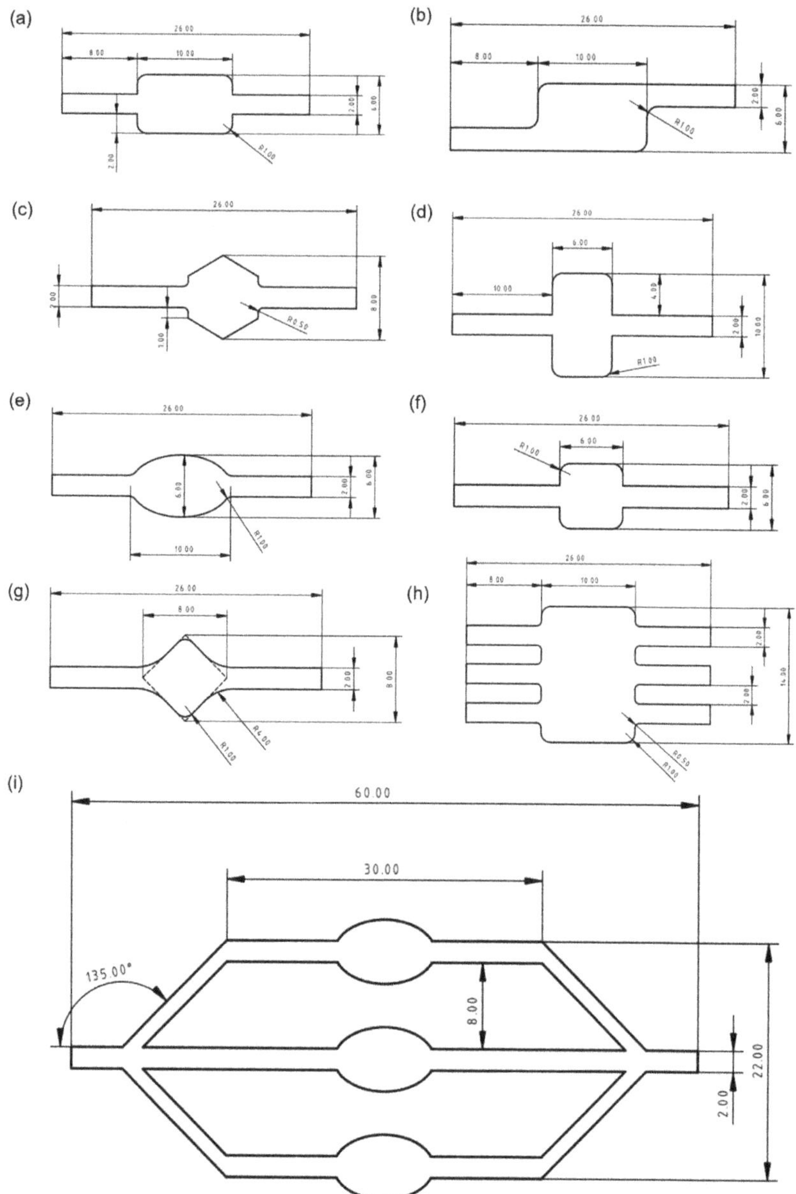

Figure 1. Studied designs. (**a**) Rectangular chamber; (**b**) Asymmetric chamber; (**c**) Wide rectangular chamber; (**d**) Hexagonal chamber; (**e**) Oval chamber; (**f**) Square chamber; (**g**) Angled square chamber; (**h**) Multiple inlets and outlets chamber; (**i**) Focus flow chamber.

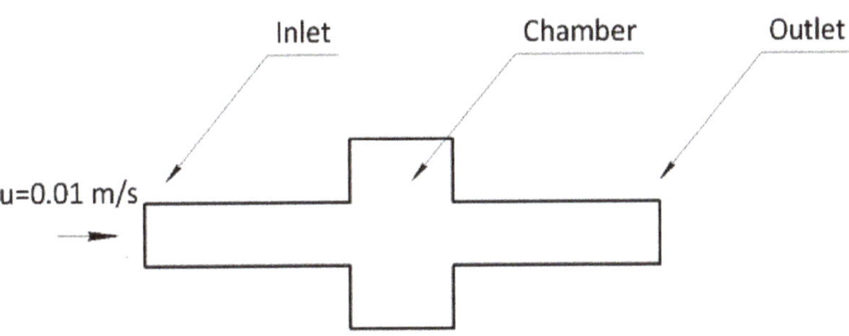

Figure 2. The physical problem.

Computational Domain and Boundary Conditions

The fluid flow is governed by the system of the Reynolds-Averaged Navier–Stokes partial differential equations and can be solved numerically using the finite element method [6–8]. In this work, the commercial software COMSOL Multiphysics was utilised in order to analyse the flow and obtain data regarding the distribution of the flow parameters inside the microfluidics circuit. To achieve this, the Laminar flow module with a Stationary time-independent study was selected. Inlet boundary condition (BC) was assigned to the respective inlet, Outlet BC was assigned to the outlet, while all remaining faces were auto-assigned with Wall BC. For the flow analysis, the computational domain was discretised using a mesh which was automatically generated and optimised for providing accuracy. The results of simulations were post-processed and visualised using the COMSOL Multiphysics postprocessor.

3. Results and Discussion

The laminar flow inside a circular pipe can be described analytically. The flow rate is linked to the pressure drop inside the pipe through the following equation:

$$Q = \Delta p (\pi R^4)/(8\eta l) \tag{1}$$

where Q is the fluid flow rate, Δp is the pressure drop between inlet and outlet of the channel, R is the channel radius, η is the fluid dynamic viscosity and l is the channel length.

The radial velocity distribution of the fully developed flow inside the cylindrical channel is described as:

$$u(r) = \Delta p\ (R^2 - r^2)/(4\eta l) \tag{2}$$

where r is the radial distance from the axis.

In order to validate the results of the selected mathematical model, a two-dimensional axisymmetric case of a 235 mm long and 100 um diameter cylindrical channel was simulated and compared with the analytical results of Equations (1) and (2). Inlet, outlet and wall boundary conditions were used in the domain boundaries. A fully developed flow was imposed at the inlet where different static pressure values were fixed from 500 mbar to 2000 mbar with a step of 500 mbar. Results from the parametric simulation are presented in Figure 3. An excellent agreement between the computed and analytical flow rate–static pressure relation (Equation (1)) and of the velocity profiles (Equation (2)). The velocity profiles show a tendency to overestimate the analytical values near the axis of the channel and the difference is greater as the pressure drop increases. These results validate the accuracy of the selected computational model in this range of laminar flow conditions inside microchannels.

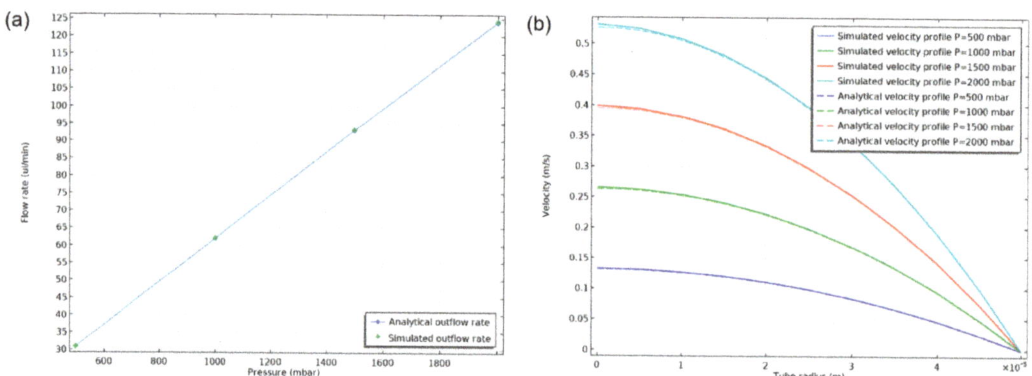

Figure 3. Comparison study results: (**a**) Relationship between the values of analytical pressure drop and flow rate against simulated values; (**b**) Comparison of analytical velocity profiles against simulated values.

One of the most important characteristics of the microfluidic chamber is its ability to maintain a homogenous flow. Meshes that were used for simulations were generated by COMSOL automatically and are shown in Figure 4. Designs were compared based on velocity magnitude inside the chamber, Figure 5. More cyan colouring of the chamber means a more even distribution of the analyte. Square and wide rectangular chambers showed large difference between the middle of the chamber and its side walls, thus making them less effective than other designs. The oval-shaped chamber presented one of the most homogenous flows, while the commonly used rectangular chamber showed good results but had some flow homogeneity around sharp corners. Asymmetric designs had a good behaviour which could be enhanced by optimising the design of the corners. The angled square resulted in a worse distribution than the hexagonal chamber design, while the latter showed relatively even distribution. The design with three inlets and outlets had one of the best distributions overall. The focus flow chamber design appeared to have a better distribution at the central chamber, while side chambers had a little less homogenous flow. Streamlines of the analyte as it moves inside the chambers can be visualised in Figure 6, and velocity vectors are presented in Figure 7. In addition, static pressure variation plots are presented in Figure 8 and show an even distribution of static pressure inside all chambers aside from the asymmetric one. Interestingly enough, it has a slightly lower relative value of pressure inside its chamber than other designs.

A useful flow quantity which can be used to assess the tendency of deformation that can be caused to the analyte molecules, is the shear rate which is calculated from the velocity gradients that are present inside the chamber. The shear rate distribution is presented in Figure 9 and shows that the asymmetric chamber, the oval chamber and its focus flow derivative are a good selection for moving the analyte though the microfluidic system without significant angular deformation effects. The hexagonal chamber, on the other hand, seems to have a distribution of high shear rate inside its domain.

In order to quantitively assess each design, the following parameters were compared in Table 1:

- Average shear rate in the chamber;
- Average pressure in the chamber;
- Pumping power.

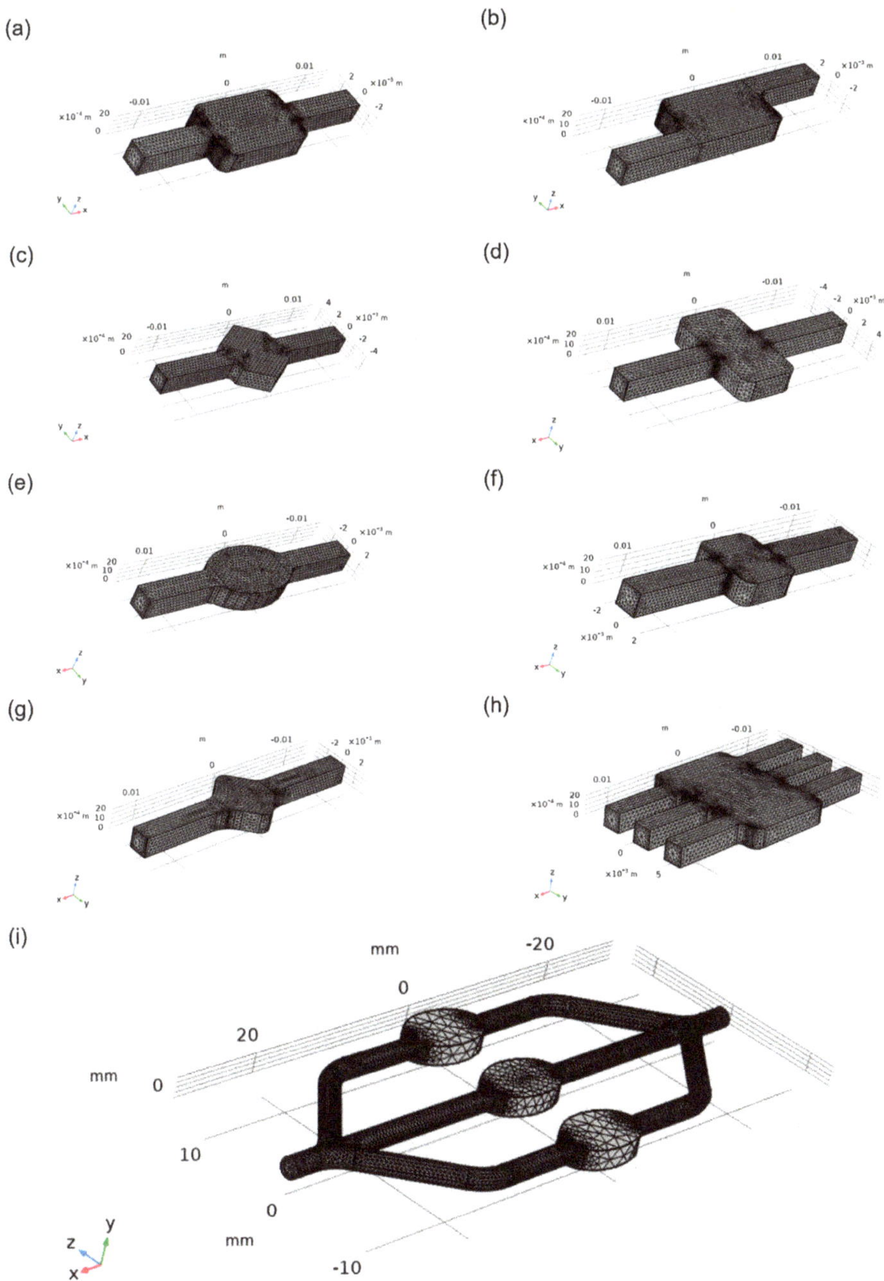

Figure 4. Generated meshes. (**a**) Rectangular chamber; (**b**) Asymmetric chamber; (**c**) Wide rectangular chamber; (**d**) Hexagonal chamber; (**e**) Oval chamber; (**f**) Square chamber; (**g**) Angled square chamber; (**h**) Multiple inlets and outlets chamber; (**i**) Focus flow chamber.

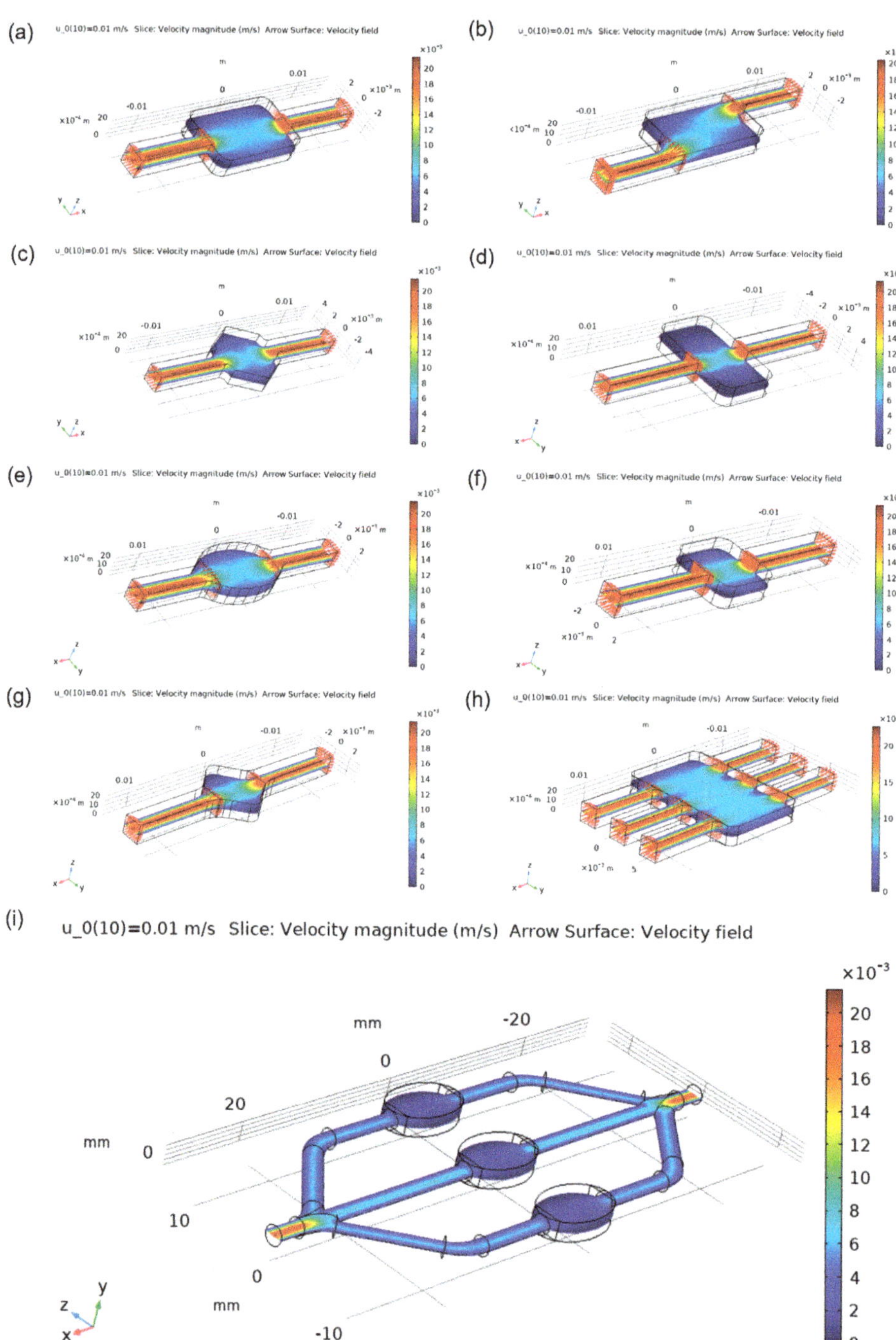

Figure 5. Velocity magnitude: (**a**) Rectangular chamber; (**b**) Asymmetric chamber; (**c**) Wide rectangular chamber; (**d**) Hexagonal chamber; (**e**) Oval chamber; (**f**) Square chamber; (**g**) Angled square chamber; (**h**) Multiple inlets and outlets chamber; (**i**) Focus flow chamber.

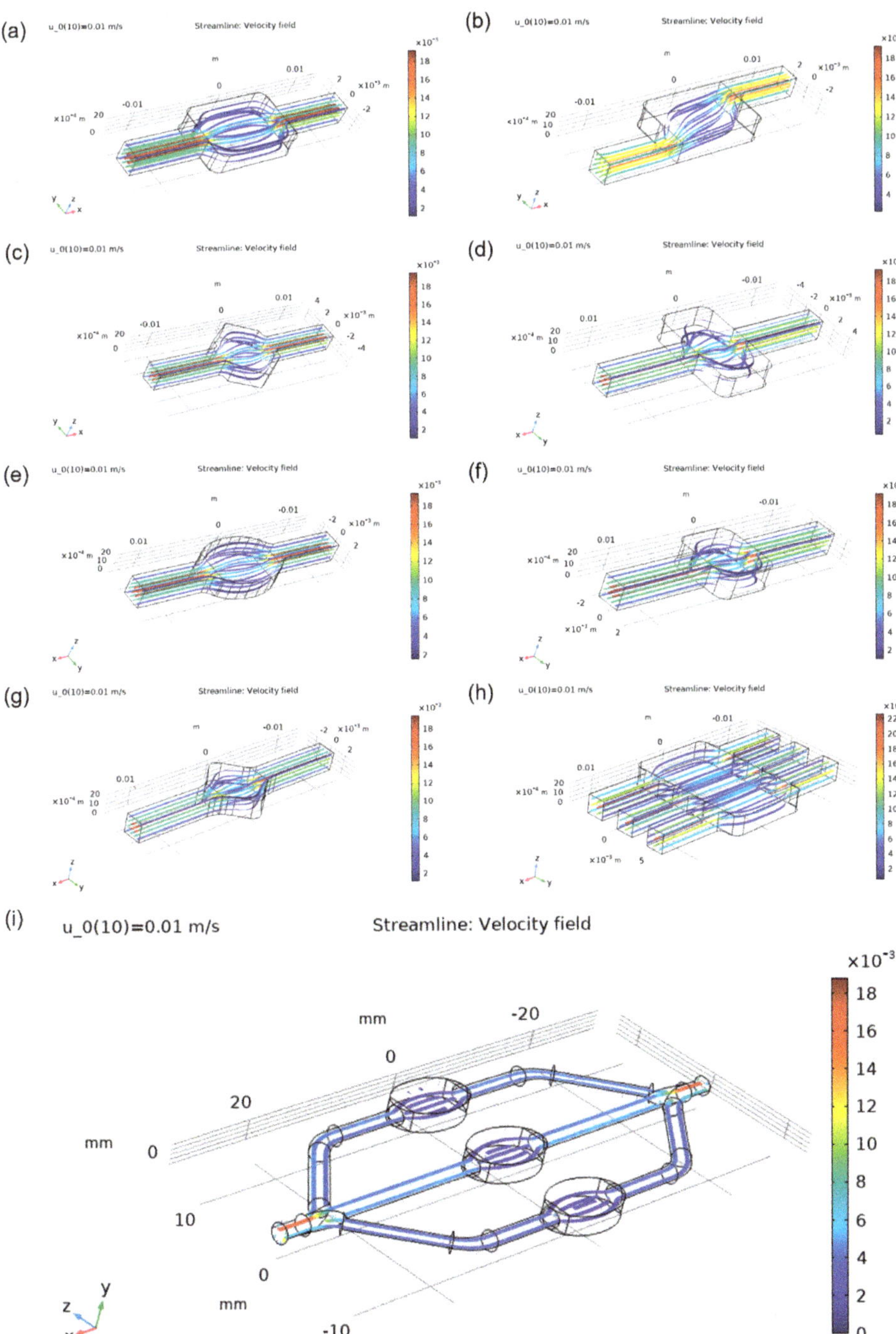

Figure 6. Streamlines: (**a**) Rectangular chamber; (**b**) Asymmetric chamber; (**c**) Wide rectangular chamber; (**d**) Hexagonal chamber; (**e**) Oval chamber; (**f**) Square chamber; (**g**) Angled square chamber; (**h**) Multiple inlets and outlets chamber; (**i**) Focus flow chamber.

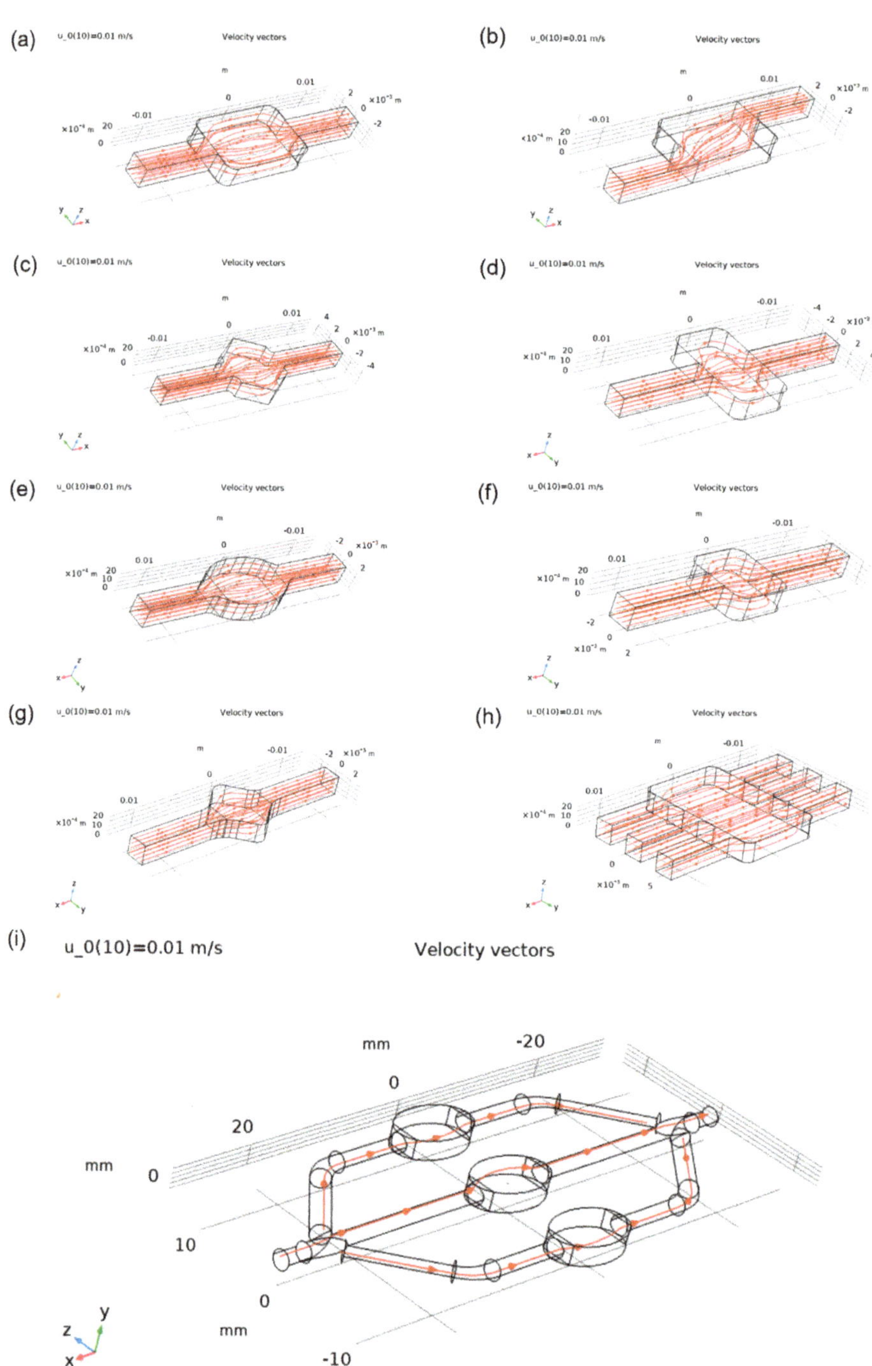

Figure 7. Velocity vectors: (**a**) Rectangular chamber; (**b**) Asymmetric chamber; (**c**) Wide rectangular chamber; (**d**) Hexagonal chamber; (**e**) Oval chamber; (**f**) Square chamber; (**g**) Angled square chamber; (**h**) Multiple inlets and outlets chamber; (**i**) Focus flow chamber.

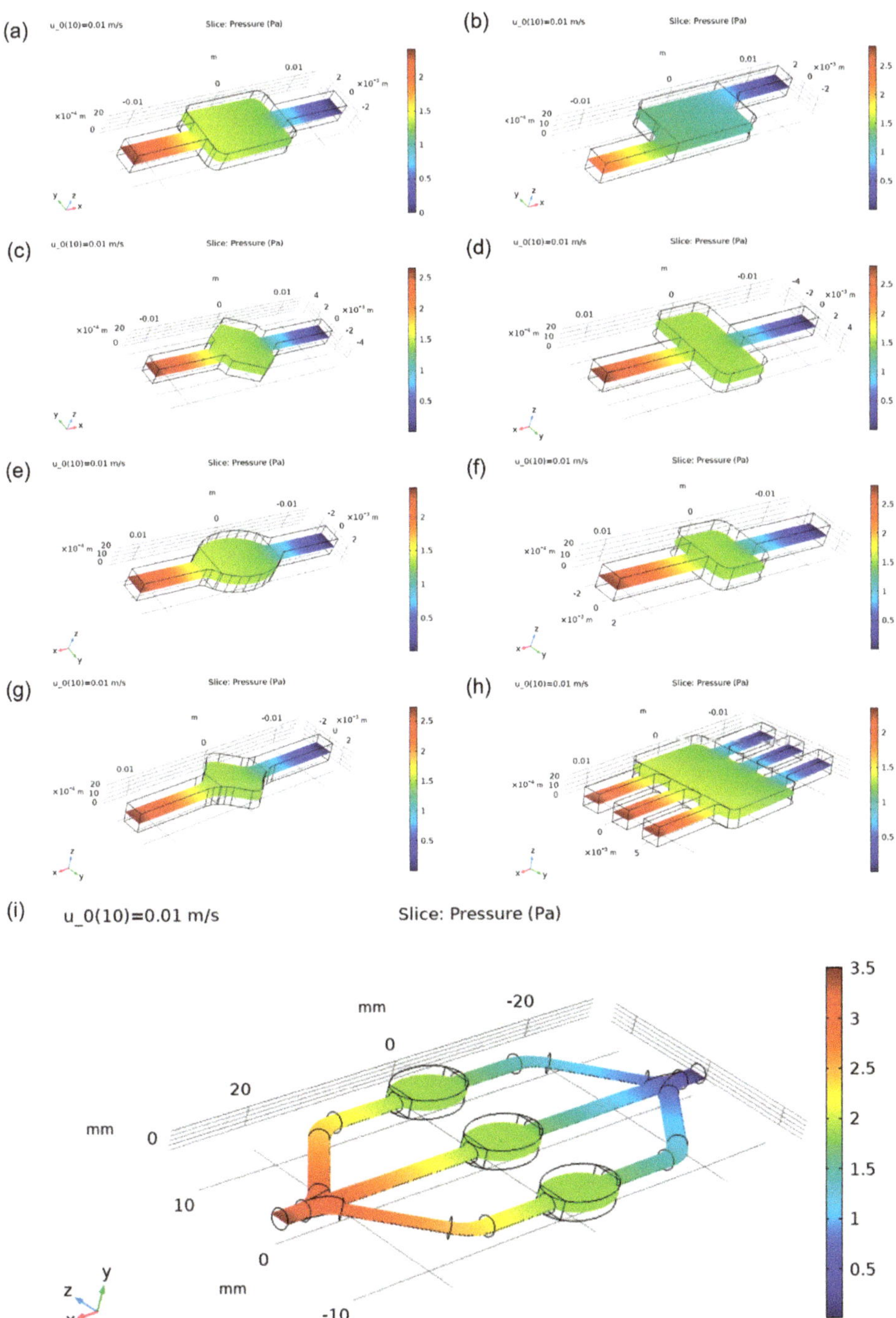

Figure 8. Pressure distribution: (**a**) Rectangular chamber; (**b**) Asymmetric chamber; (**c**) Wide rectangular chamber; (**d**) Hexagonal chamber; (**e**) Oval chamber; (**f**) Square chamber; (**g**) Angled square chamber; (**h**) Multiple inlets and outlets chamber; (**i**) Focus flow chamber.

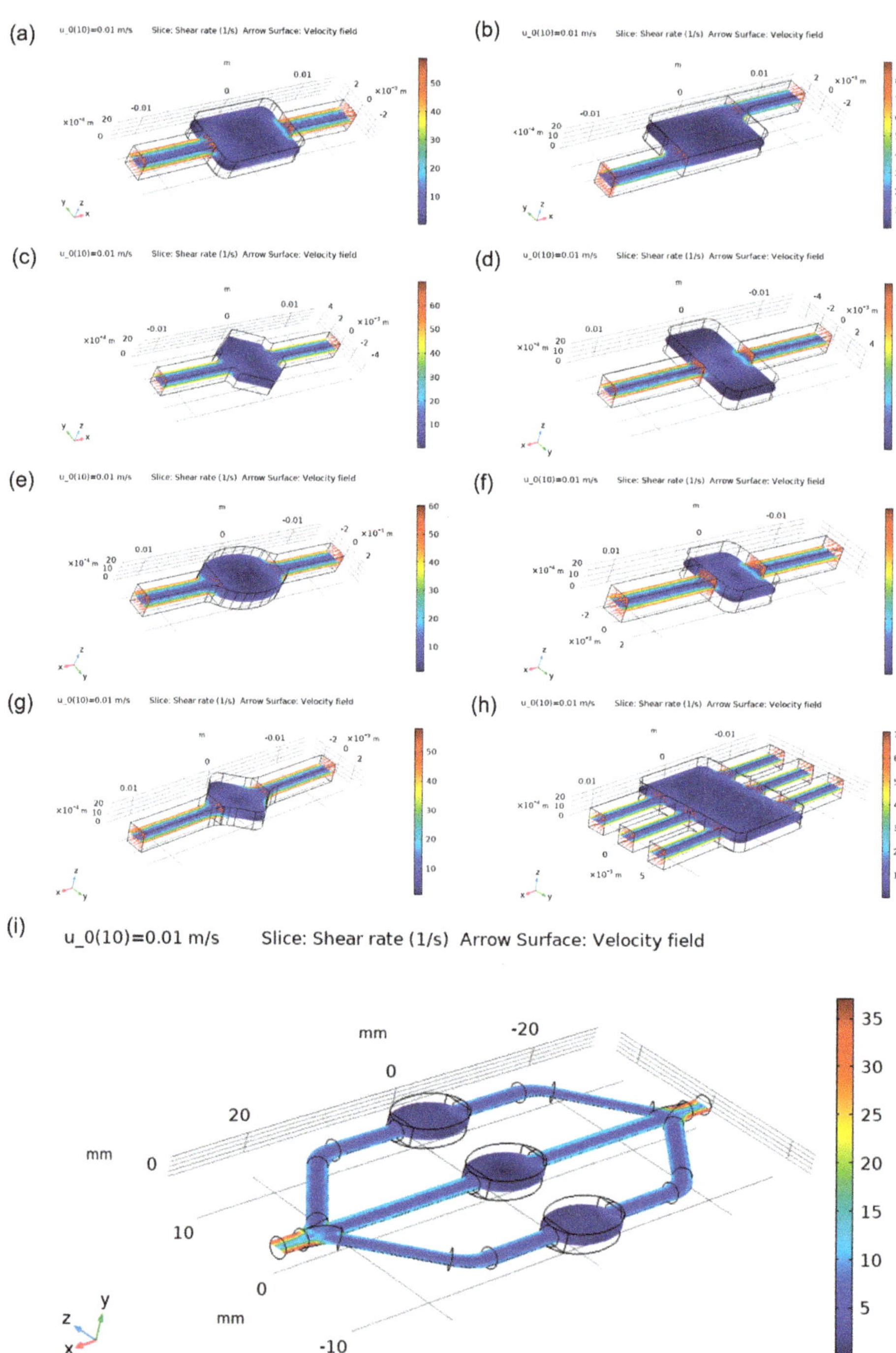

Figure 9. Shear rate: (**a**) Rectangular chamber; (**b**) Asymmetric chamber; (**c**) Wide rectangular chamber; (**d**) Hexagonal chamber; (**e**) Oval chamber; (**f**) Square chamber; (**g**) Angled square chamber; (**h**) Multiple inlets and outlets chamber; (**i**) Focus flow chamber.

Table 1. Simulation results.

Design Variant	Shear Rate, 1/s	Average Pressure in Chamber, Pa	Pumping Power, nW
Rectangular chamber	7.0423	1.2978	97.31
Asymmetric chamber	6.8807	1.2312	125.58
Hexagonal chamber	15.349	1.3771	106.67
Wide rectangle	5.0043	1.5001	113.12
Focus flow chamber	2.0822	1.875	107.12
Oval chamber	7.4903	1.315	98.283
Square chamber	8.0235	1.4972	113.45
Angled square chamber	8.6925	1.4536	110.13
Focus flow chamber	8.02	1.2672	283.94

The hexagonal chamber demonstrated the highest value of shear rate (15.349 1/s) in relation to other designs, Figure 10. Consequently, the design with multiple inlets and outlets demonstrated an average value of shear that was similar to square and oval chambers and was slightly worse than the angled square chamber. The focus flow chamber design was the least favourable choice in terms of shear rate value. Surprisingly, the rectangular chamber and its asymmetric design were almost equivalent, with shear rates of 7.0423 1/s and 6.8807 1/s, respectively.

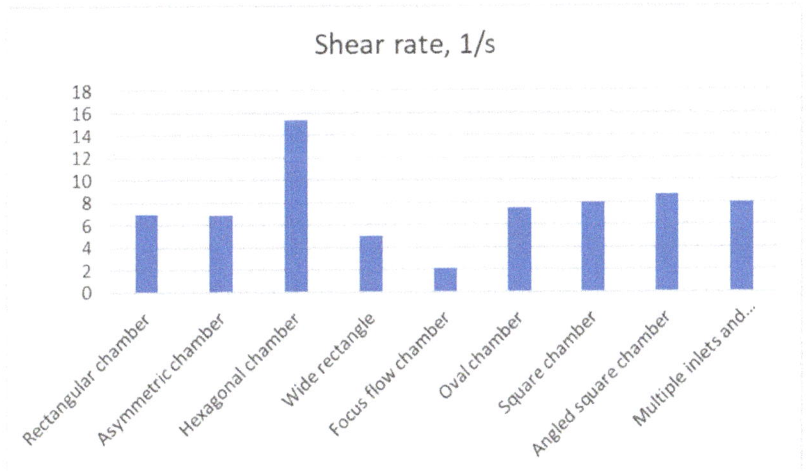

Figure 10. Average shear rate across chambers.

Due to the relatively large dimensions of the focus flow chamber design, an increased average pressure was present across all its chambers, Figure 11. On the other hand, other designs showed similar results. Approximately 1.5 Pa was registered in the square, angled square and wide chambers. Oval and rectangular chambers showed almost identical results of 1.3 Pa while the lowest result was attributed to asymmetric chamber. The hexagonal chamber had the lowest pressure from all other designs.

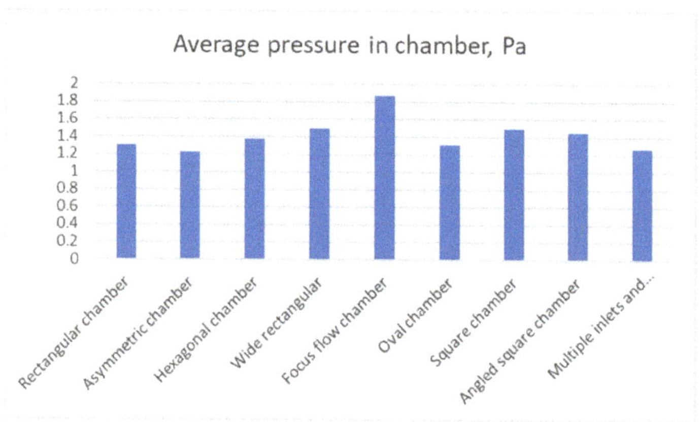

Figure 11. Average pressure in chamber.

Lastly, the mechanical power that is required by a mechanical pump in order to sustain the specified flow rate, is compared in Figure 12. The power, N, is calculated by the following relation:

$$W = \Delta p \times Q,$$ (3)

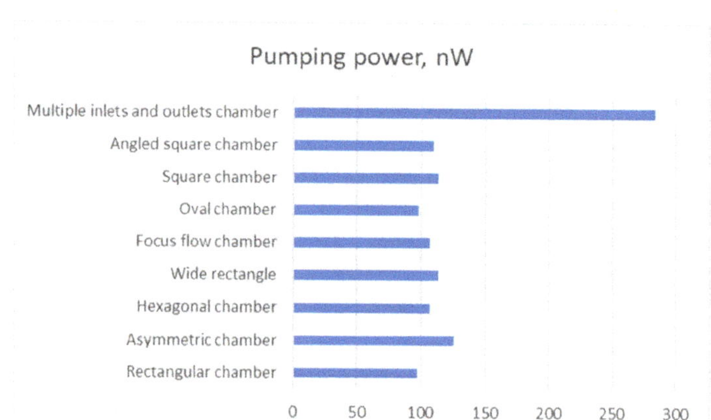

Figure 12. Mechanical pumping power.

From the results, it can be observed that the multiple inlets and outlets design consumed much more power than other designs. In fact, it consumed slightly less power than the three rectangular chambers. The second power demanding design is the asymmetric chamber at 125.58 nW. Again, the wide, square and angled square chambers show similar characteristics. The rectangular chamber requires approximately 1 nW less power than the oval chamber. Surprisingly, the focus flow chamber design required the least amount of power (82.63 nW).

In order to determine the effect of inlet flow rate on the shear rate and the pressure drop, a parametric study was undertaken by changing the inlet flow velocity. The inlet velocity changed from 0.001 m/s to 0.01 m/s with a step of 0.001 m/s. As can be deduced from the results in Figure 13, both the shear rate and pressure drop changed linearly for all the studied designs as is anticipated in this type of laminar flows.

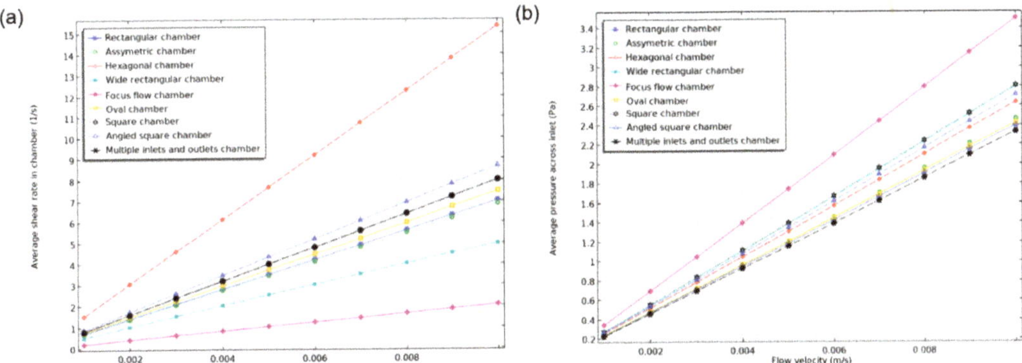

Figure 13. Comparative assessment of performance: (**a**) Relationship between the flow velocity and average shear rate in chamber; (**b**) Relationship between the flow velocity and the pressure drop.

The chambers that were designed and simulated chambers were 3D printed using Ultimaker S5 and black PLA material, Figure 14. These designs are intended to be assessed in conjunction with an Elveflow Microfluidics kit which consists of an 8 bar Jun-Air air compressor, OB1 flow controller, water reservoir, flow-rate sensor and tubing, as shown in Figure 15.

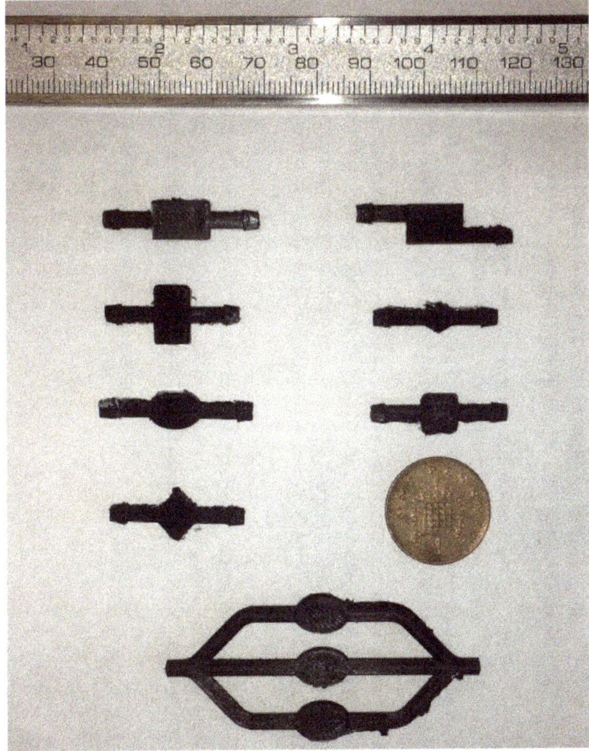

Figure 14. 3D printed chambers.

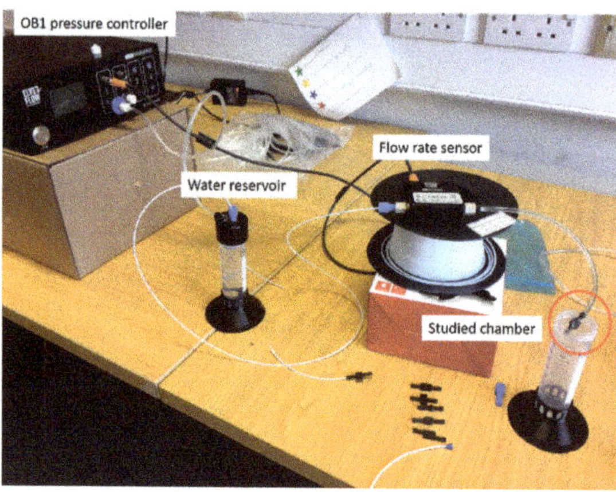

Figure 15. Experimental setup.

4. Conclusions

In this work, seven chamber designs, a focus flow chamber design and a design with three inlets and outlets were investigated. The oval-shaped chamber possessed the best velocity profile, making it a good overall choice for the microfluidics. However, the commonly used rectangular shape had the lowest pumping power requirement of 97.31 nW and the lowest pressure of 1.2978 Pa. The hexagonal shape demonstrated the highest value of shear rate at 15.349 1/s out of all designs. The focus flow chamber design showed a good scaling of the results that were obtained by the oval chamber. Yet, the middle chamber had the most homogenous flow out of three. The best velocity profile was attributed to the three inlets design at the cost of 283.94 nW, which is almost three times more than the rectangular chamber.

Author Contributions: Conceptualization, methodology, A.T. and S.D.P.; computational investigation, data analysis, I.S. and A.T.; writing—original draft preparation, visualization, I.S. and S.D.P.; writing—review and editing, supervision, project administration, funding acquisition, S.D.P. All authors have read and agreed to the published version of the manuscript.

Funding: This research received no external funding.

Data Availability Statement: Not applicable.

Conflicts of Interest: The authors declare no conflict of interest.

References

1. Teles, F.R.R.; Fonseca, L.P. Trends in DNA biosensors. *Talanta* **2008**, *77*, 606–623. [CrossRef]
2. Luka, G.; Ahmadi, A.; Najjaran, H.; Alocilja, E.; DeRosa, M.; Wolthers, K.; Malki, A.; Aziz, H.; Althani, A.; Hoorfar, M. Microfluidics Integrated Biosensors: A Leading Technology towards Lab-on-a-Chip and Sensing Applications. *Sensors* **2015**, *15*, 30011–30031. [CrossRef] [PubMed]
3. Gao, J.; Sin, M.L.; Liu, T.; Gau, V.; Liao, J.C.; Wong, P.K. Hybrid electrokinetic manipulation in high-conductivity media. *Lab Chip* **2011**, *11*, 1770–1775. [CrossRef] [PubMed]
4. Rackus, D.; Shamsi, M.; Wheeler, A. Electrochemistry, biosensors and microfluidics: A convergence of fields. *Chem. Soc. Rev.* **2015**, *44*, 5320–5340. [CrossRef] [PubMed]
5. Khan, N.I.; Song, E. Lab-on-a-Chip Systems for Aptamer-Based Biosensing. *Micromachines* **2020**, *11*, 220. [CrossRef] [PubMed]
6. Acharya, A.; Packirisamy, M. Microchambers flow simulation for immunoassay-based biosensing applications. *Photon. North* **2007**, *6796*, 679628.
7. Islamov, M.; Sypabekova, M.; Kanayeva, D.; Rojas-Solórzano, L. CFD Modeling of Chamber Filling in a Micro-Biosensor for Protein Detection. *Biosensors* **2017**, *7*, 45. [CrossRef] [PubMed]
8. Tan, S.J.; Lao, I.K.; Ji, H.M.; Agarwal, A.; Balasubramanian, N.; Kwong, D.L. Microfluidic design for bio-sample delivery to silicon nanowire biosensor—A simulation study. *J. Phys. Conf. Ser.* **2006**, *34*, 626. [CrossRef]

Proceeding Paper

Development of an All-Carbon Electrochemical Biosensor on a Flexible Substrate for the Sensitive Detection of Glucose [†]

Nikitas Melios [1], Vasiliki Tsouti [2], Stavros Chatzandroulis [2] and George Tsekenis [1,*]

[1] Biomedical Research Foundation of the Academy of Athens, 11527 Athens, Greece; nmelios@bioacademy.gr
[2] Institute of Nanoscience and Nanotechnology, NCSR "Demokritos", Patriarhou Gregoriou & Neapoleos, 15310 Aghia Paraskevi, Greece; v.tsouti@inn.demokritos.gr (V.T.); s.chatzandroulis@inn.demokritos.gr (S.C.)
* Correspondence: gtsekenis@bioacademy.gr; Tel.: +30-210-659-7476
† Presented at the 2nd International Electronic Conference on Biosensors, 14–18 February 2022; Available online: https://sciforum.net/event/IECB2022.

Abstract: Wearable biosensors for the detection of analytes in sweat are an emerging and promising technology with important applications in monitoring a person's physiological state. Sweat, being an easily accessible biofluid, shows great potential as a biological fluid for wearable devices, but also a number of challenges that must be addressed before a sensor can be commercialized. As an example, sensor fabrication on flexible substrates can greatly affect the performance of the device. Herein, the development of an enzymatic electrochemical sensor on a flexible substrate for glucose detection is presented. The sensor's three-electrode systems were made entirely with carbon-based ink on polyimide substrates and decorated with carbon black. The developed and optimized sensor design exhibited a stable and reproducible performance and was able to detect glucose in concentrations relevant to the ones present in sweat.

Keywords: glucose monitoring; electrochemical biosensor; flexible substrate; carbon black

Citation: Melios, N.; Tsouti, V.; Chatzandroulis, S.; Tsekenis, G. Development of an All-Carbon Electrochemical Biosensor on a Flexible Substrate for the Sensitive Detection of Glucose. *Eng. Proc.* **2022**, *16*, 17. https://doi.org/10.3390/IECB2022-12273

Academic Editors: Giovanna Marrazza and Sara Tombelli

Published: 14 February 2022

Publisher's Note: MDPI stays neutral with regard to jurisdictional claims in published maps and institutional affiliations.

Copyright: © 2022 by the authors. Licensee MDPI, Basel, Switzerland. This article is an open access article distributed under the terms and conditions of the Creative Commons Attribution (CC BY) license (https://creativecommons.org/licenses/by/4.0/).

1. Introduction

Recently, the increased interest in wearable or implantable sensing devices and platforms has prompted the development of highly sensitive and stable electrochemical sensors on flexible substrates [1,2]. Fabrication of these sensors demands the use of flexible substrate materials that can be bent and mounted on non-flat surfaces, such as parts of the human body. Towards this goal, polymer films such as polyethylene terephthalate (PET), polyimide (PI), polydimethylsiloxane (PDMS), as well as paper and textiles have been used as substrates for flexible biosensors [3]. Numerous examples of such sensing devices have been published, ranging from simple test strips to patches [4–6] or tattoos [7–11] and smart textiles [12], using different biological fluids such as tears [13], sweat [14], urine [15] and saliva [16]. The most common metabolic disease targeted by the majority of the developed sensors is diabetes, due to its prevalence in the population. Diabetes is routinely monitored through the detection of glucose in the bloodstream, a procedure that involves a painful and invasive blood sampling method by finger pricking [17]. Wearable devices, on the other hand, permit non-invasive sampling and continuous glucose monitoring without interrupting the wearer's daily activity [18,19] using sweat as the biological sample [20,21].

Herein, we demonstrate the fabrication of an enzymatic electrochemical sensor that has been developed on flexible substrates. Polyimide (PI) sheets have been chosen as substrates for the drawing of carbon electrodes using conductive ink [22–25]. The electrodes are made using only conductive carbon ink, while carbon black, a cost-effective carbon-based nanomaterial [26], has been applied on the working electrode. The fabricated flexible electrodes exhibit reproducible electrochemical behaviour, while the developed sensor is able to detect glucose in the concentration range of clinical significance in human sweat.

2. Materials and Methods

Conductive carbon black (Vulcan XC 27R) was kindly provided by Cabot Corporation's representatives in Greece (RAWCHEM), having an average particle size of 50 nm and a typical bulk density of 6 lbs/ft3. Carbon/graphite conductive ink C2130814D2 was purchased from Sun Chemical (Slough, UK). All other chemicals and reagents were purchased from the Sigma Chemical Company (St. Louis, MO, USA). A BioLogic SP-200 potentiostat was used for all electrochemical measurements.

2.1. Fabrication of the Enzymatic Electrochemical Glucose Sensor

PI sheets were used as the flexible substrate onto which electrodes were drawn with the use of a metal mask to imprint the electrode pattern and subsequently filled up with conductive carbon ink. The working electrodes were then further modified by placing a drop (3 μL) f carbon black suspension in an acidified aqueous solution of chitosan (0.05% in 0.05 M HCl) and were left to dry at room temperature. This task was performed twice before moving on to the immobilization of the enzyme. For the latter, a 2 μL drop of glucose oxidase (40 mg/mL) was added and was also left to dry at room temperature. Finally, a 2 μL drop of Nafion (0.1% in H_2O) was placed on top to complete the sensor fabrication process.

2.2. Sensor Characterization

Chronoamperometric detection was used for glucose detection and quantification. Measurements were performed at an applied voltage of 0.6 V, and the current was recorded for 180 s so that a stable value was obtained. A range of different concentrations of glucose in PBS 10 mM pH 7.4 was tested by placing a drop on the surface of the sensor so that all three electrodes were covered.

3. Results and Discussion

3.1. Flexible Electrodes Fabrication and Performance

Polyimide (PI), polyethylene terephthalate (PET), cyclic olefin copolymer (COC), and polydimethylsiloxane (PDMS) were tested as substrate materials for the fabrication of the glucose sensor due to their flexibility. Electrodes fabricated on PI sheets showed the best adhesion to the substrate, and this is why PI was chosen for the fabrication of the sensor. A three-electrode system was hand-drawn on the PI substrate using a carbon/graphite ink. To achieve the same pattern each time, a stainless-steel mask was fabricated with the electrode design. The mask was used to carve the design on the PI sheets before drawing the electrodes. The hand-drawn electrodes were reproducibly fabricated with very small performance variation. This was validated by cyclic voltammetry measurements (Figure 1). To increase the sensitivity of the working electrode, carbon black was used as a cost-efficient alternative to other carbon-based nanomaterials, such as carbon nanotubes and graphene oxide, which have been used for the fabrication of electrochemical glucose sensors [27–29]. Carbon black exhibits excellent electrical conductivity and offers fast electron kinetics due to its numerous defect sites, thus making it an ideal candidate for the development of low-cost biosensors [26].

Figure 1. Hand-drawn all-carbon three electrode system on PI sheets.

3.2. Enzymatic Glucose Sensor Performance

The working electrode was further modified with glucose oxidase, which was adsorbed on top of a chitosan layer and stabilized with the use of Nafion. Glucose oxidase entrapment allowed the enzyme to maintain its activity, whereas crosslinking the enzyme with glutaraldehyde had an impact on the enzyme's active site and its turnover number and thus decreased the linear dynamic range over which glucose could be detected and also increased the limit of detection (results not shown). A one-step modification was also tested by mixing the enzyme with the carbon black/chitosan dispersion and depositing it on the surface of the working electrode. Once again, the performance of the sensor was inferior to that of the sensor, where the enzyme was adsorbed on the chitosan matrix and stabilized by Nafion. The obtained results from the chronoamperometric interrogation of the sensor (Figure 2) demonstrate that the sensor was able to detect glucose at concentrations ranging from 0.05 mM to 2 mM (Figure 3). Despite, therefore, the simple design of the sensor, glucose could be detected at concentrations relevant to sweat and within the range of clinical significance [14,30,31]. Lower limits of detection have been achieved in other published work, albeit with the use of mediator-modified electrodes such as Prussian blue [5], or with much more complex designs involving numerous steps for the biomodification of the working electrode.

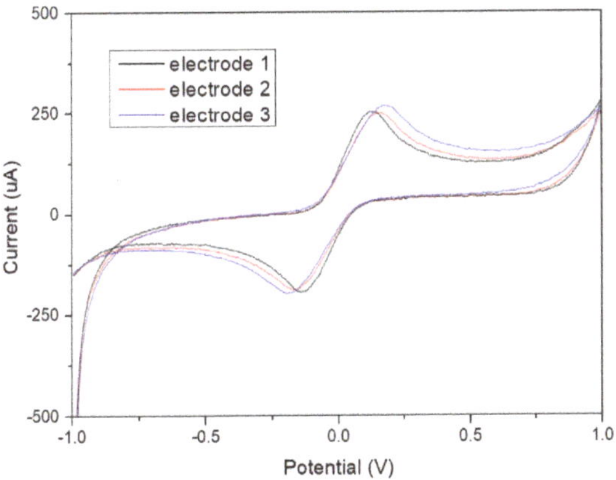

Figure 2. Cyclic voltammogram of 10 mM ferrocyanide/ferrocyanide redox couple in PBS 1x pH 7.4.

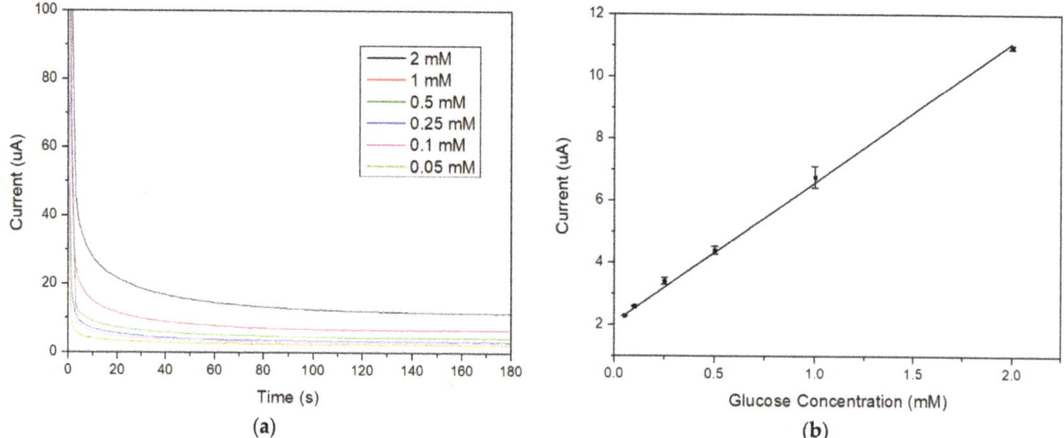

Figure 3. (**a**) Chronoamperometric response of the biosensor to increasing glucose concentrations from 0 to 2 mmol L^{-1} in PBS 1x pH 7.4. (**b**) Corresponding calibration curve. The points in the plot are the mean value $\pm$ SD ($n = 3$).

4. Conclusions

Flexible electrochemical sensors are the key to the development of wearable devices that could offer continuous monitoring of biomarkers in relevant biological fluids. Herein, we have demonstrated the sensitive and reproducible detection of glucose with the use of a hand-drawn three-electrode system fabricated entirely with the use of carbon ink. Furthermore, the utilization of carbon black allowed us to achieve a very low limit of detection and a linear dynamic range at glucose concentrations relevant to human sweat. The sensor could be further integrated with a microfluidic network to develop a wearable patch for glucose monitoring or could be used as the basis for the development of sensors for other biomarkers of clinical or metabolic significance in sweat.

Author Contributions: Conceptualization, G.T., V.T. and S.C.; methodology, G.T.; validation, N.M.; writing—original draft preparation, N.M.; writing—review and editing, G.T.; visualization, super-vision, G.T., V.T. and S.C.; project administration, G.T. All authors have read and agreed to the published version of the manuscript.

Funding: This research received no external funding.

Conflicts of Interest: The authors declare no conflict of interest.

References

1. Yoon, J.; Cho, H.Y.; Shin, M.; Choi, H.K.; Lee, T.; Choi, J.W. Flexible electrochemical biosensors for healthcare monitoring. *J. Mater. Chem. B* **2020**, *8*, 7303–7318. [CrossRef] [PubMed]
2. Yang, A.; Yan, F. Flexible electrochemical biosensors for health monitoring. *ACS Appl. Electron. Mater.* **2021**, *3*, 53–67. [CrossRef]
3. Wang, S.; Chinnasamy, T.; Lifson, M.; Inci, F.; Demicri, U. Flexible substrate-based devices for point-of-care diagnostics. *Trends Biotechnol.* **2016**, *34*, 909–921. [CrossRef] [PubMed]
4. Martín, A.; Kim, J.; Kurniawan, J.F.; Sempionatto, J.R.; Moreto, J.R.; Tang, G.; Campbell, A.S.; Shin, A.; Lee, M.Y.; Liu, X.; et al. Epidermal Microfluidic Electrochemical Detection System: Enhanced Sweat Sampling and Metabolite Detection. *ACS Sens.* **2017**, *2*, 1860–1868. [CrossRef]
5. Wiorek, A.; Parrilla, M.; Cuartero, M.; Crespo, G.A. Epidermal Patch with Glucose Biosensor: PH and Temperature Correction toward More Accurate Sweat Analysis during Sport Practice. *Anal. Chem.* **2020**, *92*, 10153–10161. [CrossRef]
6. Cho, E.; Mohammadifar, M.; Choi, S. A single-use, self-powered, paper-based sensor patch for detection of exercise-induced hypoglycemia. *Micromachines* **2017**, *8*, 265. [CrossRef]
7. Emaminejad, S.; Gao, W.; Wu, E.; Davies, Z.A.; Nyein, H.Y.Y.; Challa, S.; Ryan, S.P.; Fahad, H.M.; Chen, K.; Shahpar, Z.; et al. Autonomous sweat extraction and analysis applied to cystic fibrosis and glucose monitoring using a fully integrated wearable platform. *Proc. Natl. Acad. Sci. USA* **2017**, *114*, 4625–4630. [CrossRef]

8. Kim, J.; Sempionatto, J.R.; Imani, S.; Hartel, M.C.; Barfidokht, A.; Tang, G.; Campbell, A.S.; Mercier, P.P.; Wang, J. Simultaneous Monitoring of Sweat and Interstitial Fluid Using a Single Wearable Biosensor Platform. *Adv. Sci.* **2018**, *5*, 1800880. [CrossRef]
9. Pu, Z.; Su, X.; Yu, H.; Li, D. Differential Sodion-Based Self-Calibrated Epidermal Microfluidic System for Continuous Glucose Monitoring. In Proceedings of the IEEE 32nd International Conference on Micro Electro Mechanical Systems (MEMS), Seoul, Korea, 27–31 January 2019; pp. 429–432. [CrossRef]
10. Yokus, M.A.; Songkakul, T.; Pozdin, V.A.; Bozkurt, A.; Daniele, M.A. Wearable multiplexed biosensor system toward continuous monitoring of metabolites. *Biosens. Bioelectron.* **2020**, *153*, 112038. [CrossRef]
11. Yoon, S.; Yoon, H.; Ko, S.; Park, C.; Zahed, M.A.; Park, J. A Flexible Electrochemical-Physiological Epidermal Hybrid Patch for Chronic Disease Management. In Proceedings of the IEEE Sensors, Rotterdam, The Netherlands, 25–28 October 2020; pp. 1–4. [CrossRef]
12. He, W.; Wang, C.; Wang, H.; Jian, M.; Lu, W.; Liang, X.; Zhang, X.; Yang, F.; Zhang, Y. Integrated textile sensor patch for real-time and multiplex sweat analysis. *Sci. Adv.* **2019**, *5*, eaax0649. [CrossRef]
13. Park, J.; Kim, J.; Kim, S.-Y.; Cheong, W.H.; Jang, J.; Park, Y.-G.; Na, K.; Kim, Y.-T.; Jun, H.H.; Lee, C.Y.; et al. Soft, smart contact lenses with integrations of wireless circuits, glucose sensors, and displays. *Sci. Adv.* **2018**, *4*, eaap9841. [CrossRef]
14. Lee, H.; Song, C.; Hong, Y.S.; Kim, M.S.; Cho, H.R.; Kang, T.; Shin, K.; Choi, S.H.; Hyeon, T.; Kim, D.H. Wearable/disposable sweat-based glucose monitoring device with multistage transdermal drug delivery module. *Sci. Adv.* **2017**, *3*, e1601314. [CrossRef]
15. Comer, J.P. Semiquantitative Specific Test Paper for Glucose in Urine. *Anal. Chem.* **1956**, *28*, 1748–1750. [CrossRef]
16. Valdés-Ramírez, G.; Bandodkar, A.J.; Jia, W.; Martinez, A.G.; Julian, R.; Mercier, P.; Wang, J. Non-invasive mouthguard biosensor for continuous salivary monitoring of metabolites. *Analyst* **2014**, *139*, 1632–1636. [CrossRef]
17. Kim, J.; Campbell, A.S.; de Ávila, B.E.F.; Wang, J. Wearable biosensors for healthcare monitoring. *Nat. Biotechnol.* **2019**, *37*, 389–406. [CrossRef]
18. Sabu, C.; Henna, T.K.; Raphey, V.R.; Nivitha, K.P.; Pramod, K. Advanced biosensors for glucose and insulin. *Biosens. Bioelectron.* **2019**, *141*, 111201. [CrossRef]
19. Teymourian, H.; Barfidokht, A.; Wang, J. Electrochemical glucose sensors in diabetes management: An updated review (2010–2020). *Chem. Soc. Rev.* **2020**, *49*, 7671–7709. [CrossRef]
20. Kim, J.; Campbell, A.S.; Wang, J. Wearable non-invasive epidermal glucose sensors: A review. *Talanta* **2018**, *177*, 163–170. [CrossRef]
21. Jadoon, S.; Karim, S.; Akram, M.R.; Kalsoom Khan, A.; Zia, M.A.; Siddiqi, A.R.; Murtaza, G. Recent developments in sweat analysis and its applications. *Int. J. Anal. Chem.* **2015**, *2015*, 164974. [CrossRef]
22. Bndodkar, A.J.; Jeang, W.J.; Ghaffari, R.; Rodgers, J.A. Wearable sensors for biochemical sweat analysis. *Annu. Rev. Anal. Chem.* **2019**, *12*, 1–22. [CrossRef]
23. Taleat, Z.; Khoshroo, A.; Mazloum-Ardakani, M. Screen-printed electrodes for biosensing: A review (2008–2013). *Microchim. Acta* **2014**, *181*, 865–891. [CrossRef]
24. Cinti, S.; Mazzaracchio, V.; Cacciotti, I.; Moscone, D.; Arduini, F. Carbon black-modified electrodes screen-printed onto paper towel, waxed paper and parafilm. *Sensors* **2017**, *17*, 2267. [CrossRef]
25. Karuwan, C.; Wisitsoraat, A.; Phokharatkul, D.; Sriprachuabwong, C.; Lomas, T.; Nacapricha, D.; Tuantranont, A. A disposable screen printed graphene-carbon paste electrode and its application in electrochemical sensing. *RSC Adv.* **2013**, *3*, 25792–25799. [CrossRef]
26. Arduini, F.; Cinti, S.; Mazzaracchio, V.; Scognamiglio, V.; Amine, A.; Moscone, D. Carbon black as an outstanding and affordable nanomaterial for electrochemical (bio)sensor design. *Biosens. Bioelectron.* **2020**, *156*, 112033. [CrossRef]
27. Liu, Y.; Wang, M.; Zhao, F.; Xu, Z.; Dong, S. The direct electron transfer of glucose oxidase and glucose biosensor based on carbon nanotubes/chitosan matrix. *Biosens. Bioelectron.* **2005**, *21*, 984–988. [CrossRef]
28. Kang, X.; Wang, J.; Wu, H.; Aksay, I.A.; Liu, J.; Lin, Y. Glucose Oxidase-graphene-chitosan modified electrode for direct electrochemistry and glucose sensing. *Biosens. Bioelectron.* **2009**, *25*, 901–905. [CrossRef]
29. Zheng, B.; Cheng, S.; Liu, W.; Lam, M.H.W.; Liang, H. Small organic molecules detection based on aptamer-modified gold nanoparticles-enhanced quartz crystal microbalance with dissipation biosensor. *Anal. Biochem.* **2013**, *438*, 144–149. [CrossRef]
30. Bruen, D.; Delaney, C.; Florea, L.; Diamond, D. Glucose sensing for diabetes monitoring: Recent developments. *Sensors* **2017**, *17*, 1866. [CrossRef]
31. Lee, H.; Hong, Y.J.; Baik, S.; Hyeon, T.; Kim, D.H. Enzyme-Based Glucose Sensor: From Invasive to Wearable Device. *Adv. Healthc. Mater.* **2018**, *7*, 1701150. [CrossRef]

MDPI
St. Alban-Anlage 66
4052 Basel
Switzerland
Tel. +41 61 683 77 34
Fax +41 61 302 89 18
www.mdpi.com

Engineering Proceedings Editorial Office
E-mail: engproc@mdpi.com
www.mdpi.com/journal/engproc

www.ingramcontent.com/pod-product-compliance
Lightning Source LLC
LaVergne TN
LVHW070642170726
843515LV00004B/304